COSMOLOGY

CONTENTS

INTRODUCTION: WHAT IS COSMOLOGY?

The modern definition of the science of cosmology is: *the scientific investigation that deals with the current structure, origin, evolution and future of the entire universe as a physical system*. The goal of this field of science is to create a comprehensive story – based upon our best current theories of matter, space and time – that explains the many observations which scientists have accumulated over the years. These observations have been made with the use of telescopes and other forms of sophisticated astronomical instrumentation, both ground-based and space-borne.

The advancement of cosmology is deeply connected with refinements in our understanding of the fundamental ingredients of the universe, including matter, energy, space and time. These form the core of the investigations undertaken by physicists and astronomers to discern the basic natural laws within which our universe appears to operate. This information is largely codified within the conceptual domains of Quantum Mechanics and General Relativity. The former theory describes in detail the fundamental ingredients of matter in what is called the Standard Model. The latter theory describes how gravity operates in terms of a detailed mathematical rubric in which the gravitational field is synonymous with the geometric properties of a four-dimensional spacetime continuum.

Modern cosmology had its beginnings in the work by the 17th century English scientist Sir Isaac Newton, whose meticulous mathematical description of gravity was rapidly expanded to explain the previously mysterious movements of the planets, and also used to craft the first scientific model of the cosmos operating under Universal Gravitation. Previous 'proofs' of the infinite scale of the universe were mostly philosophical or religious in nature. With Newton, there was now the first physics-based, dynamical explanation for the vastness of cosmological space. Today, a relativistic theory of gravity called General Relativity, developed by the German physicist Albert Einstein, has led to the articulation of 'Big Bang cosmology', the highly successful, mathematical framework upon which all of modern cosmology is now based.

Scientific cosmology seeks to uncover the mysteries of the universe, such as the nature of two major 'dark' components to the universe: 'dark matter' and 'dark energy'. The quest for answers is aided by parallel investigations being undertaken at major physics laboratories around the world. The recent detection of gravity waves has reaffirmed the position of Einstein's General Relativity as the premier theory of gravity. Meanwhile, astronomers continue to obtain telescopic information about the large-scale structure of the universe and its earliest history by direct imaging of the formation of the first stars and galaxies within 100 million years after the 'birth' of the universe. They also quantify and interpret the cosmic microwave background (CMB) radiation, which is now recognized as a physical medium that records the evolution of 'dark matter' clustering in the early universe, along with imprints of the earliest eras in cosmic history known as 'inflation'.

Camille Flammarion's wood engraving, coloured by a modern artist, from L'atmosphère: météorologie populaire (1886), an early work that sought to demystify astronomy for the masses.

The continued investigation of the earliest eras in the evolution of the universe has led to the application of sophisticated physical theories in so-called 'quantum gravity research' to the elucidation of the initial moments in the formation of the universe. In this arena, cosmology has largely become a sub-area of theoretical, high-energy physics and the search for a unified theory of nature. This is a highly mathematical undertaking that places cosmological 'origins' issues at the crux of a deep understanding of the nature of the physical world and reality itself: Are space and time quantized? Does a multiverse exist? A variety of astronomical observations may help theoretical physicists answer many of these very subtle, but profound, questions. There is much that we do

know, and much that we still need to discover. So for anyone thinking about studying for a degree in cosmology, there has never been a better time to learn about our amazing universe; how it started, what's happening now, and how it will all end.

The enormously large and small scales of modern cosmological research require the use of what is called scientific notation such that a number like 149 is rendered as 1.49×10^2 and 0.000657 is rendered as 6.57×10^{-4}. There will also be occasion to use prefixes such as kilo, mega, or micro to refer to some physical units such that 12,000,000 parsecs would be expressed as 12 megaparsecs (mpc) or 0.000013 meters would be expressed as 13 micrometers. Throughout this book, I will adopt the international system of units (SI). Forces will be referred to in the units of Newtons and temperature will be in the Kelvin scale or Celsius scale as needed. These will be augmented by the astronomical measures of Astronomical Units (AU) as 1.496×1011 metres, light years as 9.46×10^{12} km and parsecs as 3.26 light years.

Chapter One
DISCOVERING THE UNIVERSE

Dawn of Reason and Practical Cosmology – Abstract Thinking – Megalithic Record-keeping – Ancient Cosmologies: Babylon, India, Greece – Star Catalogues: Hipparchus, Ptolemy – Kepler's Solar System – Physical Scale of the Solar System – Galilean Telescopes

THE DAWN OF REASON AND PRACTICAL COSMOLOGY

Forty thousand years ago, our ancestors were pragmatic hunter-gatherers living on the edge of survival. They spent most of their time searching for food plants or following the migration routes of their principle food animals.

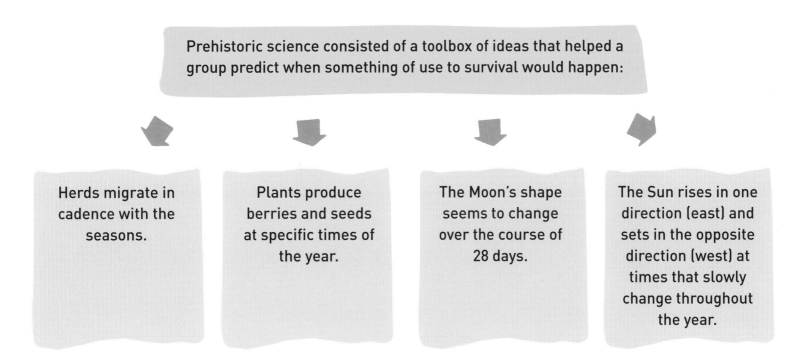

Prehistoric science consisted of a toolbox of ideas that helped a group predict when something of use to survival would happen:

Herds migrate in cadence with the seasons.

Plants produce berries and seeds at specific times of the year.

The Moon's shape seems to change over the course of 28 days.

The Sun rises in one direction (east) and sets in the opposite direction (west) at times that slowly change throughout the year.

There is a cyclical pattern to these natural movements, which our ancestors would have recognized and could have used to predict events on Earth that were important to their survival – and to plan accordingly.

The phases of the moon in photographic time-lapse.

The stars in the sky form patterns that move westward month on month, but the patterns themselves remain fixed from generation to generation. These star patterns are known to us today as *constellations*. The constellation Orion the Hunter always looks like Orion the Hunter. Scorpius the scorpion always looked like Scorpius the scorpion. And every night the entire sky seems to rotate around a fixed point in the sky which has come to be known as 'north'. The opposite direction, 'south', is the direction you should travel in the northern winter to get to warmer climates, and the direction you should travel in the northern summer to get to cooler climes. These are such basic observations that, given how our ancestors had much the same intellectual capacity as we do today, it is unimaginable that they didn't know about them or the underlying astronomical world, no matter what stories they contrived to explain them.

Cosmology and Brain Evolution

The creation of cosmological concepts requires a set of skills and capacities that derive from the very way in which our brains are put together and have evolved over the millennia. To create a stable and accurate model of the world, the first thing the brain has to do is to have a sense of its own body and how it is located in space. It also has to identify this 'self' as being different from that of other people or objects. If it cannot do this accurately, it cannot decide how to move in space, anticipate the consequences of that movement, or how to anticipate and empathize with the actions of other people. Most of this work is handled by the temporoparietal junction (TPJ), which takes information from the limbic system (emotional state) and the thalamus (memory) and combines it with information from the visual, hearing and internal body sensory systems to create an integrated internal model of where your body is located in space. Next, the posterior cingulate body lets us experience your body as having a definite location in space, and that this location is where you, the 'Self' is located. Finally, the Posterior Superior Parietal Lobule gives us a sense of the boundary between your physical body and the rest of the world. When activity in this brain region is reduced, you experience the feeling of having 'merged with the universe' and your body is in some way infinite.

Just as the brain generalizes a collection of associations in space to define the concept of 'cat', it can detect patterns in time in the outside world and begin to see how one event leads to another as a rule of thumb or a law of nature. This perception of relationships is one of cause and effect and is due to activity in the cerebellum and the hippocampus. For humans, all of these brain regions have evolved over millions of years to allow us to experience the basic components of the objective physical world, extract logical relationships from it, and from these elements fashion cosmology as a subject of inquiry.

ABSTRACT THINKING

Much of prehistoric science is only a minor extension of the basic knowledge shared by migratory animals, which certainly are aware of seasonal and diurnal changes and time their migrations accordingly. But our ancestors took a further step. They came up with the remarkable idea of communicating and recording their knowledge in various ways and passing it on to later generations. The most dramatic means, shared to some degree by our earlier *Homo neanderthalensis* predecessors, was to draw on cave walls 65,000 years ago. But in addition to accurate portrayals of the various animals that were important to them, there were also ladder-like shapes, dots and handprints painted and stencilled deep within the Aviones and Maltravieso Caves in Spain.

Neanderthal Art from Blombos Cave, South Africa about 73,000 years ago.

Zigzag markings have been identified on a shell found in Indonesia dating back 500,000 years and thought to be the work of another early human species, *Homo erectus*. A piece of red ochre carved with similar zigzag lines, found at Blombos Cave in South Africa, has been dated to about 73,000 years ago. By 40,000 years ago, we have observations of nature and natural cycles as a precursor to the contemporary scientific method of investigating nature, and the recording of this information in various ways as a precursor to mathematics, becoming an increasing pre-occupation among our prehistoric ancestors, even to the extent that they included in their artwork many abstract and non-representational images.

A replica of the Blanchard bone, a lunar calendar made by the Aurignacian culture approximately 32,000 years ago in the Bordeaux region.

In addition to the motion of the Sun, Moon, stars and planets, there were other details of the sky with which they were no doubt familiar. Our distant ancestors were able to see much the same night sky that you and I do when we travel far from city lights. The dramatic Milky Way and its faint haze of light cuts across the sky at an angle very different from the Sun's apparent path through the sky (known as the *Ecliptic plane*), and is a very easy and commanding target for the eye. If, after that initial glimpse of the sky, you begin to study its details more carefully, other objects begin to stand out that are not merely pinpoint stars.

A night-time photograph of the Milky Way.

ECLIPTIC PLANE ▶ *the apparent path of the Sun through the sky.*

The Pleiades star cluster – also known as the 'Seven Sisters' – has been noted by many civilizations in both the Old World and the New World. The Nebra Sky Disk from Northern Germany, which shows these seven stars, dates from 1600 BCE. The Babylonians called this star system *MulMul*, or star-star, in their catalogues, which date from 2300 BCE. A number of other non-stellar objects can also be easily seen with the dark-adapted naked eye, including the Andromeda Galaxy, the Great Nebula in Orion, and the Hercules star cluster. Planets, comets, meteors and the Moon are the brightest non-stellar objects visible in the sky; yet, apart from Greek mythology and the origin of the Pleiades as the daughters of Atlas, our ancestors seemed to have paid much less attention to these fuzzy, indistinct objects. That would have to wait many thousands of years until human technology allowed us to investigate these objects and deduce from them the true scale of the universe.

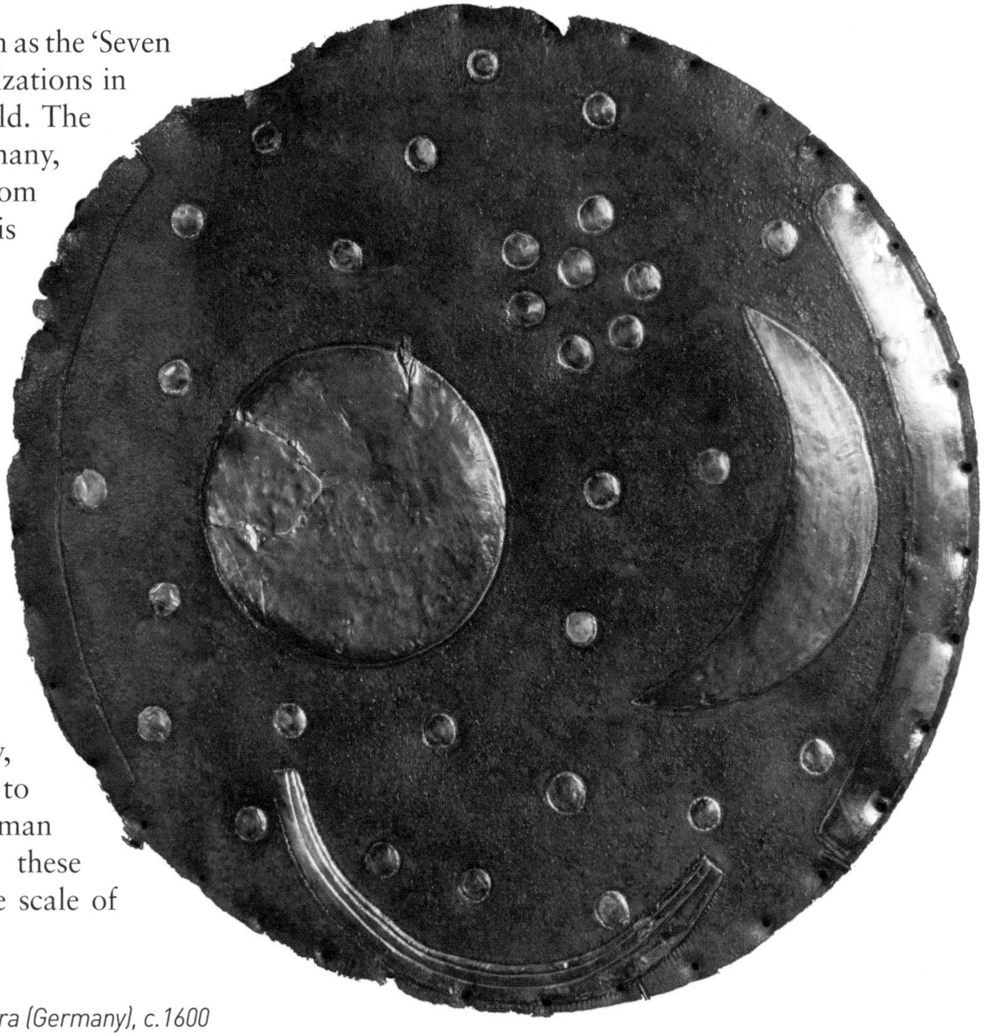

The Nebra Sky disk, found at Mittelberg near Nebra (Germany), c.1600 BCE. It is interpreted generally as a Sun or full Moon, a lunar crescent and stars (including a cluster interpreted as the Pleiades). Two golden arcs along the sides, marking the angle between the solstices, were added later.

MEGALITHIC RECORD-KEEPING

The most impressive example of our ancestors' practical interest in matters of astronomy appears in the form of massive monuments of stone with alignments related to the basic cycles of the Sun and Moon.

The most famous of these is Stonehenge in England, which was constructed in several stages between 3000 BCE and 2000 BCE. At the summer solstice, the Sun rises over the Heel Stone as viewed from the centre of the monument. An even clearer alignment, this time with the winter solstice, can be found in the Newgrange burial mound in Ireland, which was probably erected c.3200 BCE. Here, the inner chamber is illuminated via a narrow passageway for 17 minutes precisely on the shortest day of the year.

An even earlier monument with solar alignments can be found in Nabta Playa in Egypt. This collection of stones was assembled around 4800 BCE to form a ring, and includes an alignment with the summer solstice. Other calendric or celestial alignments may also exist, such as a purported alignment with the rising of the star Sirius found by multiple investigators.

The Nabta Playa stone circle in Egypt.

The Goseck circle in Germany was constructed in 4900 BCE and so is about the same age as Nabta Playa. It consists of a concentric ditch 75 m (246 ft) across and two rings containing entrances at places aligned with sunrise and sunset on the winter solstice. Smaller entrances appear to be aligned with the summer solstice.

There is also the recently discovered Neolithic temple at Göbekli Tepe, in Turkey, dating back to around 9000 BCE. These large stones include an apparent alignment with the rising of the bright star Sirius. Due to the Earth's precession, Sirius would have made its appearance above the horizon as a 'new star' around this time, as viewed from this location. Precession refers to the Earth's 'wobble' – the change in the direction of our planet's axis – which over time brings new areas of the sky into view and hides others.

PRECESSION ▶ *the slow change in the direction of the Earth's axis of rotation, which occurs over a period of 25,772 years.*

The rising of certain stars, such as Sirius, were known to be important predictors of the annual floods of the Nile, which made the land so fertile. Ancient Egyptian monuments, including the famous Pyramid of Khufu on the Giza Plateau, were carefully aligned with the Pole Star (Thuban at that time) c.2600 BCE. The temple to Amon-Re in Karnak, c.2000 BCE, was also aligned with the rising of the winter solstice Sun, at which time the Sun's rays fell within the centre of the temple, illuminating for a few hours the sacred Holly of Hollies. This is similar to an alignment for Abu Simbal in Egypt built in 1255 BCE which occurs on October 21 and February 21, but may be designed for specific festival days or the coronation of Ramesses II himself.

Generally, stars were curiosities in the sky. Although no extant prehistoric records of a stellar universe are available from most locations around the world, we find in ancient Egypt tomb paintings and papyrus records that date from about 2100 BCE. In particular, the so-called Diagonal Star Tables, frequently inscribed in coffin lids, revealed the Egyptian constellations. These 36 constellations (*decans*) sequentially rose above the eastern horizon after sunset every ten days and their brightest stars were listed in a haphazard manner. Elaborate New Kingdom paintings such as those found on the tomb ceilings of Senemut (1473 BCE) and Ramesses IV (1100 BCE) bear witness to the fact that the starry sky played some role in their mythology.

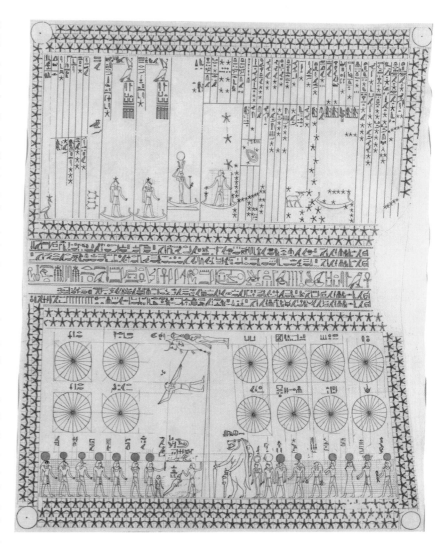

The constellations, represented in Senenmut's Tomb, 1473 BCE.

By 1500 BCE, the cuneiform clay-tablet records of Babylonian and Sumerian astrologers were not only acknowledging the existence of planets, especially Venus, but had established a more or less fixed collection of constellations in their mythology. The modern Zodiac is largely a construct of the Babylonians. Meanwhile, ancient Chinese astrologers had extensively followed sunspots and solar eclipses as part of their own divination methods.

ZODIAC ▶ *a belt of sky that encompasses the apparent paths of the Sun, Moon and visible planets. It is divided into 12 regions, each one named after the constellation it holds.*

ANCIENT COSMOLOGIES: BABYLON, INDIA, GREECE

On the Indian sub-continent, the origins of astronomy and astrology can be traced back to around 2000 BCE. Much of what we know about Indian astronomy comes from the Sanskrit sacred books called the *Vedas*. Vague references to the Sun being at the centre of the universe exist in Vedic writings from as early as 3000 BCE. There was substantial interest in measuring the heavens and detecting mathematical regularities in the movements of the planets that led to the development of Indian astrology at around the same time as their Babylonian contemporaries. By the 6th century BCE, one of the Vedic schools, the Vaisheshika, professed an early atomistic view of nature in which the four Aristotelian elements: Earth, Air, Fire and Water, were enlarged to nine: Earth, Air, Fire, Water, Aether, Time, Space, Soul and Mind. The idea that time and space were reducible to their own atoms was a truly unique perspective, not to be revisited until the mid-1900s.

Indian cosmology is also unique in that it offered far more quantitative detail to the structure and changes in the universe; far beyond what Babylonian or ancient Greek stories could provide. According to Hindu Vedic cosmology, there is no absolute start to time. Time is considered infinite and cyclic. Similarly, the universe has neither a beginning nor an end but is cyclical. The current universe is just the start of the present cycle. Each cycle is a period of one day in the life of Brahma, and lasts 8.6 billion years. A Brahman Year is over 1 trillion human years. Brahma lives for 100 of his years before all worlds and souls are completely dissolved for all eternity!

Ancient Greek measuring devices aided the foundation of the first critical study of the stars.

The constellation Orion is an unmistakable pattern of stars known for millennia.

The rise of Greek civilization c.900 BCE led to the first documented ideas about stars and the celestial universe. Among the earliest subjects mentioned were eclipses, the Pleiades star cluster, the constellation Orion and the bright star Sirius c.700 BCE. Anaximander and Philolaus made extensive references to stars, planets and a model for the planetary system, with the Earth or some other unseen body at the centre of these motions. Democritus even proposed that the bright band on the night sky, the Milky Way, might consist of distant stars.

Plato, Eudoxis, and Aristotle developed in some detail the idea that the objects in the sky were affixed to concentric 'celestial spheres' nested one within the other. Beyond the planetary spheres was the sidereal sphere, where stars 'hung' like lanterns. Within the Earth–Moon region existed everything that changed, including such transitory things as thunderstorms, rainbows, meteors and comets. Everything beyond the Moon remained the same forever. Aristotle proposed the Four Elements – Earth, Air, Fire and Water – but this was extended to a fifth element called 'Quintessence' to account for the eternal and perfect substance of the stars.

Aristotle was one philosopher who developed the idea of 'celestial spheres.'

In *De Luce* (On Light), written in 1225 CE, the English theologian Robert Grosseteste even went so far as to explore the nature of matter and the cosmos. He described the birth of the universe in an explosion, and the crystallization of matter to form stars and planets in a set of nested spheres around Earth. *De Luce* is the first known attempt prior to the 17th century to describe the heavens and Earth using a single set of physical laws.

STAR CATALOGUES: HIPPARCHUS, PTOLEMY

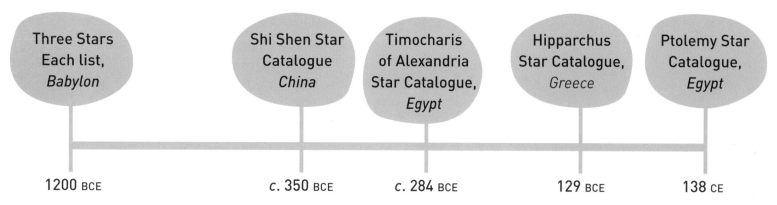

Three Stars Each list, *Babylon* — 1200 BCE

Shi Shen Star Catalogue *China* — c. 350 BCE

Timocharis of Alexandria Star Catalogue, *Egypt* — c. 284 BCE

Hipparchus Star Catalogue, *Greece* — 129 BCE

Ptolemy Star Catalogue, *Egypt* — 138 CE

The most ancient star catalogue that has survived to the present time is the 'Three Stars Each' list of the Babylonians c.1200 BCE, which consisted of nothing more than the three brightest stars in each of the 12 zodiacal houses. In the 1,000 years that separate the Babylonian and Greek periods, there is no surviving evidence of anyone using angle measures like degrees to explicitly quantify the actual locations of the stars in an accurate sky map. After 300 BCE, Timocharis, Aristarchus, Aristillus, Archimedes and Hipparchus can be counted among the first individuals to have used degree measures by dividing the circle into 360 degrees, with each degree consisting of 60 arc minutes. A star catalogue created by the Chinese astrologer Shi Shen, which included 800 entries, was developed c.350 BCE with an angle measured from the north celestial pole. The angle was a Chinese degree such that one degree represented one day of sky movement or $^{360}/_{365}$ Greek degrees.

Claudius Ptolemy

This Greco-Roman astronomer and polymath lived during the 2nd century CE and was responsible for one of the only surviving astronomical documents from these ancient times: the *Almagest*. This book contained his star catalogue based upon the more-ancient Hipparchus catalogue. It also included his geocentric, mathematical model of the solar system in tabular form based upon the planets affixed to epicycles carried on celestial spheres that rotated with Earth at its centre.

Aristarchus of Samos was one of the first Greek astronomers.

A mural from c. 1598 showing an astronomer attempting to determine a star's altitude with the help of his assistants.

From his Alexandria observatory, the Greek astronomer and philosopher Timocharis recorded that the star Spica was located 8 degrees west of the Autumnal equinox, but that is all that has survived from his labours. By 284 BCE, other records suggest he was the first western astronomer to create a star catalogue of the several-hundred brightest stars using a cross-staff. Along with theodolites, cross-staffs were used to measure the angular elevation of a star above the horizon. In the latter case, a staff with perpendicular sliding pieces was sighted towards the star and the crosspiece slid until it spanned the distance from the star to the horizon. The corresponding angle could be calculated from simple trigonometry.

An example of the use of a quadrant instrument from a 1564 illustration. attempting to determine a star's altitude with the help of his assistants.

In 129 BCE, the Greek astronomer and mathematician Hipparchus continued and surpassed the works of Timocharis by measuring the positions of over 8,000 naked-eye stars. But the Hipparchus star catalogue also vanished in antiquity, and his only surviving work is his *Commentary on the Phaenomena of Aratus and Eudoxus*, which describes the celestial constellation figures in detail. The accuracy of these angle measurements was about the diameter of the full moon, or 0.5 degrees.

KEPLER'S SOLAR SYSTEM

Circa 1580, the Danish astronomer Tycho Brahe developed specialized instruments – quadrant circles and sextants. These gave positions for the celestial bodies which were ten times more accurate than those of Ptolemy or Hipparchus. Tycho calculated the positions of 1,000 stars, with accuracies of $1/_{60}$ degree, which formed the basis for the *Rudolphine Tables* developed by his assistant Johannes Kepler.

Johannes Kepler

Born on 27 December 1571 in Germany, Kepler went to work for Tycho Brahe as his mathematical assistant on 4 February 1600 with the assigned goal of making sense of the voluminous data being accumulated by Brahe, and specifically to prove Brahe's hybrid solar system model in which the planets orbited the sun but the sun orbited Earth. Meanwhile, Kepler succeeded in discovering the elliptical shape of the orbit of Mars and uncovered what would be called 'Kepler's Three Laws of Planetary Motion'. While not performing these calculations, he was an accomplished and famous court astrologer, and in 1621 was called upon to defend his mother against a charge of witchcraft.

Johannes Kepler was known for his laws on planetary motion.

Most importantly, Tycho observed a comet in 1585 and through careful measurements worked out its path among the stars. It was clear from contemporaneous measurements made by the German astronomer and mathematician Michael Maestlin that this object showed no evidence of a shift in its position when measured from two different locations (known as parallax). This meant it must be an object located far beyond the 'corruptible and changeable' sub-lunar spheres. Its path was not circular, but intersected the orbits of the planets. One could conclude from this only that these celestial spheres were not at all solid, but far more ephemeral – if they existed at all. As Tycho put it '...the structure of the universe is very fluid and simple'.

PARALLAX ▶ *the apparent shift in position of a celestial object when viewed from two different locations on Earth or from the same location on Earth six months apart. The degree of shift can then be used by means of geometry to calculate the distance to the object.*

Careful studies of the planetary positions measured by Tycho allowed his assistant Johannes Kepler to identify three 'laws' that were seemingly obeyed by the five known planets:

- Planets move along elliptical paths.
- Planetary speeds are such that they sweep out equal orbit areas in equal times.
- The orbit periods are related to their solar distances by the rule $T^2 = R^3$, where T is the orbital period of a planet and R is the average orbital radius.

1st Law

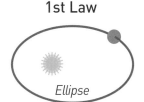

Ellipse

2nd Law

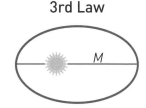

S_2

S_1

*Equal area in the
same time area*
S_1 = area S_2

3rd Law

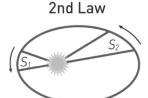

M

*P: period (the time for one cycle)
M: length of the major axis
p^2/M^3 is the same for all planets*

ASTRONOMICAL UNIT (AU) ▶ *the distance between the Earth and the Sun; a relative scale that compares distances between the Sun and other solar system bodies.*

Kepler's 'Third Law' set the scale for the entire solar system. By setting the Earth–Sun distance as 1.0 units (called *Astronomical Units* or AUs), Mercury would be at 0.35 AU and Saturn would be at 9 AU. For the first time in history, the relative scale of the solar system was established. Its absolute dimensions in kilometres only required that one of the planetary distances be determined.

An example of a solar system model with nested shells c.1540.

Parallax

The subtended angle denoted by the Greek letter θ is measured in radian units such that 1 radian equals 57.3⁰, which obtains from the ratio 360/2π.

$$\theta = 57.3 \frac{Displacement\ (m)}{Distance\ (m)}$$

The parallax angle, measured in degrees, obtains when you view an object of a fixed diameter or separation from a distance. This effect can also be reversed so that if you measure the parallax angle with a theodolite, and know beforehand the size of the distant object in metres, you can determine its distance also in metres. In astronomy, we measure objects in the sky in degrees or in arcminutes ($^1/_{60}$ degree), arcseconds ($^1/_{3600}$ degree) or even smaller units. When multiplied by their distance using the formula above, we can determine the physical size of distant objects in space, an important clue to their natures.

*Galileo Galilei
developed the refracting telescope.*

PHYSICAL SCALE OF THE SOLAR SYSTEM

Although Hipparchus pioneered the method of parallax to gauge the distance to the Moon, when parallax or other methods based upon lunar phases (Aristarchus) were applied to the Earth–Sun distance, the results were inaccurate by factors of a thousand. Ptolemy's estimate of 1,200 Earth radii, or about seven million km (four million miles), remained the gold standard for the AU until the time of Nicolaus Copernicus, the Polish Renaissance astronomer and mathematician whose revolutionary model of the universe placed the Sun rather than the Earth at its centre.

A much more precise value was finally determined for the first time using telescopic techniques by Jean Richer and Giovanni Cassini. They measured the parallax of Mars between Paris and Cayenne in French Guiana when Mars was at its closest to Earth in 1672. From their measured parallax angle, and the predicted distance to Mars based on Kepler's AU scale, they deduced that 1 AU corresponded to a distance of 21,700 Earth radii, or 140 million km. The adopted modern value is 149 million km.

GALILEAN TELESCOPES

Developments in the refracting telescope in the early 17th century led to a major change in understanding of the true content of the sky and the universe. Many objects such as the Pleiades star cluster, and the nebulosities in Orion and Andromeda, had been known since antiquity, but never in enough detail to discern their true shapes, or whether they were the only such objects in the sky. In 1610, the Italian astronomer Galileo Galilei, using his own pioneering refracting telescope design, revealed that the Milky Way was full of stars too faint for the naked eye to see. This was a major theological calamity, for why would God have created invisible stars? Galileo did not bother to mention the Orion nebula, though he drew a map of the stars within it, but his work set the stage for other astronomers to build larger instruments and begin the long process of cataloguing the contents of the universe.

A woodcut of the telescope used by Johannes Hevelius from 1673.

William Herschel built larger telescopes, which helped him to discover Uranus.

By 1655, the Dutch astronomer Christiaan Huygens was constructing even larger 'spy glasses', but these unwieldy telescopes were rapidly eclipsed by telescopes based upon Isaac Newton's mirror design, known as reflecting telescopes, beginning in 1668. The mirrors were disks of metal called specula that were ground so that the concave shape was a section of a sphere or a paraboloid. This allowed the light from a distant object to be focused, and by using a second lens system called the eyepiece, astronomers obtained a greatly magnified image. The Anglo-German astronomer Sir William Herschel improved on the 'Newtonian' design to build increasingly larger telescopes. The first of these, completed in 1789, was a 49-inch (1.2 m) instrument that he used primarily to search for double stars. An early success that made him internationally famous was the discovery of the planet Uranus, the first planetary body discovered since antiquity.

The French comet hunter Charles Messier published his 'Catalogue of Nebulae and Star Clusters' in 1781. This featured some 103 objects, including the Orion Nebula (Messier 42), the Andromeda Nebula (Messier 31) and the Crab Nebula (Messier 1). He was not particularly interested in the objects themselves. His aim was to ensure they would not be mistaken for comets by serious comet hunters. This was a rather haphazard list obtained through a small telescope. However, between 1782 and 1802, Herschel used several telescopes of 30-cm (12-in) and 45-cm (18-in) apertures to conduct the first systematic searches for non-stellar objects in the sky. He ultimately catalogued 2,400 objects that he called 'nebulae', which included faint star clusters among many other 'smudges' of light. These were collected and enlarged by the work of his sister Caroline Herschel and his son John Herschel into the New General Catalogue of 7,840 deep-sky objects. This led to the modern usage of 'NGC' to identify many of the bright nebulae, star clusters and extra-galactic nebulae studied today. The Orion Nebula can either be called Messier 42 (M 42) or NGC 1976, for example. John Herschel later took the 601 cm (20-ft) telescope to Cape Town, South Africa, where he catalogued the stars of the southern skies.

In 1845, William Parsons, the 3rd Earl of Ross, completed an even larger telescope with a 72-in (183-cm) aperture. With this he was the first to discern the spiral shapes of several nebulae, in particular the Whirlpool Nebula (M 51).

The 40-foot (12-metre) telescope used by William Herschel to catalogue nebulae.

SPACE AND COMMON SENSE?

The issue of whether space was empty or filled has been debated many times throughout history. Some ancient philosophers, such as Aristotle, assumed that the pure empty space of a void was impossible, as *'nature abhors a vacuum'*. Atomists such as Democritus thought otherwise, however. If matter was made from small particles (atoms), then there had to be gaps between them so that matter could move around – otherwise, like a traffic jam on a motorway, everything would be permanently locked in place at one location.

The Pythagoreans, too, believed that a vacuum could exist, while Parmenides attacked this idea. Those that believed empty space was impossible were forced by their own logic to propose something to fill up the vacuum. Anaximander, of the Ionian School in Greece, believed that everything in the universe was composed of a single substance that was a 'continuous and infinite medium' filling all space: the *Aether*.

The 17th-century French scientist René Descartes did not believe that empty space could exist. He argued that extension is the fundamental property of matter, and that extension without matter was impossible. He said that *'a vacuum of space in which there is absolutely no body is repugnant to reason.'* This led Descartes to the idea that space must be completely filled by a rarefied medium that could not be detected by the senses. He proposed that there must be three kinds of matter:

- Fire was the substance of the stars and the Sun, consisting of minute, luminous particles;
- Air was made from transparent spherical particles through which light could pass; *and*
- Earth was the substance from which all the planets were made.

The need for a medium to fill space was not abandoned until the 20th century, when the last candidate, the Luminiferous Ether, was dismissed during the post-Einstein revolution offered by Special and General Relativity. General Relativity finally explained how space itself was a fiction because the Newtonian concept of an Absolute Space and Time framework for the cosmos was no longer needed, and was not consistent with the confirming observations of Relativity.

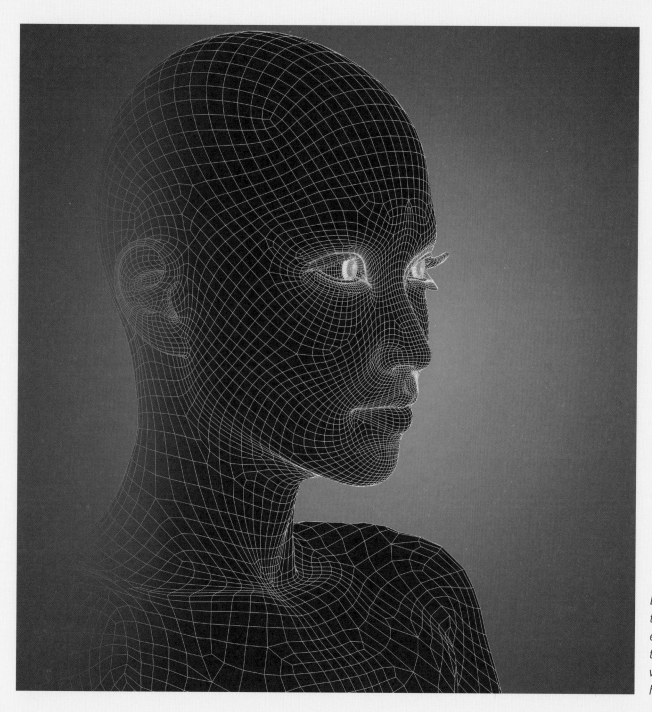

Earlier ideas claimed that space could not exist without a body to define it – as in this wireframe model of a human head.

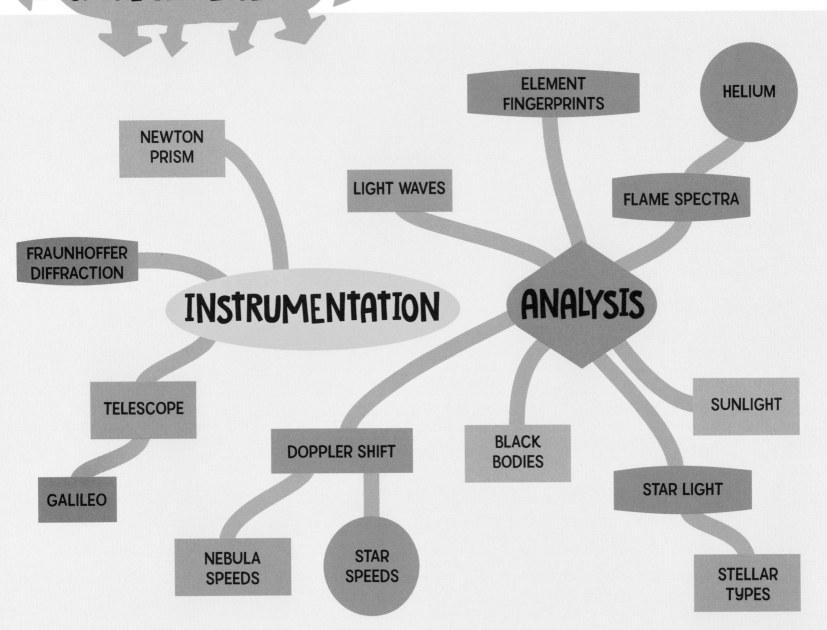

THE BIRTH OF ASTROPHYSICS

In 1835, Auguste Comte, a prominent French philosopher, stated that humans would never be able to understand the chemical composition of stars. He was soon proved wrong...

'...we would never know how to study by any means [the stars'] chemical composition...I persist in the opinion that every notion of the true mean temperatures of the stars will necessarily always be concealed from us.'
— Auguste Comte

Astrophysics is the analysis of celestial objects and phenomena in terms of the fundamental laws of physics, including Newton's laws of motion and gravitational mechanics, the laws and principles of electrodynamics developed in the 1800s, and the many new physical theories developed during the 20th century.

Astronomer Simon Newcomb's *Elements of Astronomy*, published in 1900, defines the field of astrophysics as having started with the invention of spectroscopy in 1859. *Spectroscopy* has enabled astronomers to measure the spectrum of electromagnetic radiation that radiates from celestial objects including the stars. From this, scientists can deduce important information such as the stars' chemical compositions and their relative speeds and directions of movement in relation to the Earth and other stars. Spectroscopy was reinforced by the use of photography in astronomy, heralded by the first clear photo of the Moon by John William Draper in 1840.

The ancient Romans were the first to discover that a prism could separate sunlight into colours. A variety of investigators in the 1600s, such as the Anglo-Irish chemist Robert Boyle, also experimented with this effect. But it is Sir Isaac Newton who is credited with the first recorded, detailed study of this process in his book *Optiks*. Moreover, he discovered that a second prism could recombine these colours into white light. So, prisms were not merely 'colourizing' light but dramatically displaying an inherent property of light itself.

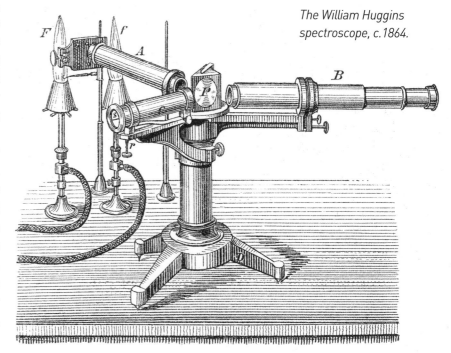

The William Huggins spectroscope, c.1864.

In 1802, William Wollaston developed a prismatic system that included a lens to focus sunlight on a screen, discovering that the rainbow of colours was interrupted by innumerable dark lines of varying width and intensity. Ten years later, a Bavarian lensmaker, Joseph von Fraunhofer, improved Wollaston's system by replacing the prism with a number of parallel rectangular slits formed from 260 parallel wires to diffract the light into a spectrum. This led to the creation of ruled gratings with thousands of slit-like lines scored into optical glass, known as a *diffraction grating*. The principle was that the light from a source passes through a *collimating lens* (which creates parallel light rays), and then through a slit and on to the surface of a grating, where *optical interference* dispersed the light into a spread of wavelengths. A small telescope could then be added at the angle specified by the grating geometry and focused for the eye or on to photographic emulsion to record.

The spectrum of the Sun showing atomic absorption lines, plotted by Joseph Fraunhofer, ordered from letters A to K from longest to shortest wavelengths.

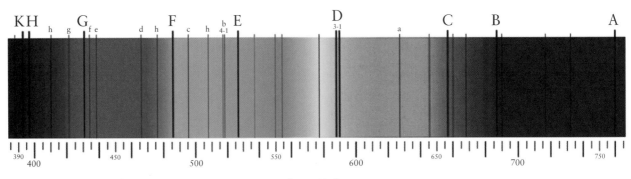

wavelength in mm

The spectroscope of the Fraunhofer design was quickly applied to study light from a variety of sources. In 1859, German chemist Robert Bunsen and physicist Gustav Kirchhoff applied this instrument to studying the light from various elements heated by candle flames. They discovered that the spectral lines served as 'fingerprints'; their distinctive and unique patterns could be used with high fidelity to identify the chemical elements. This realization led almost immediately to the addition of this instrument to telescopes to analyze the light from planets, stars, the Sun, and distant nebulae; a technique first employed by the Italian astronomers Giovanni Donati and Angelo Secchi, and the American Lewis Rutherfurd.

The spectra of some common elements showing their different 'fingerprints'.

COMPOSITION AND CLASSIFICATION OF STARS

Towards the end of the 19th century, scientists such as William Huggins and Angelo Secchi gathered as many star spectra as possible, and spent a considerable amount of time placing them into a variety of classification schemes. Three basic groups emerged: blue and white stars, yellow (or solar-type stars), and red stars. In 1885, Edward Pickering, of the Harvard College Observatory, aided by his team of female 'computers' including Williamina Fleming and Annie Jump Cannon, undertook an ambitious programme of stellar spectral classification using spectra recorded on photographic plates. By 1890, a catalogue of over 10,000 stars had been prepared and grouped into thirteen spectral types. By 1924, Annie Jump Cannon, following Pickering's vision, had expanded the catalogue to nine volumes and over a quarter of a million stars, and developed a system of seven spectral types – O, B, A, F, G, K, M – which astronomers quickly accepted for worldwide use.

NAMES TO KNOW: INVESTIGATING LIGHT

Sir Isaac Newton (1643–1727)

William Wollaston (1659–1724)

Joseph von Fraunhofer (1787–1826)

Robert Bunsen (1811–1899)

Gustav Kirchoff (1824–1887)

SPECTRAL TYPE	COLOUR	TEMPERATURE (K)	SPECTRAL FEATURES
O	Dark blue	28,000–50,000	Ionized helium
B	Medium Blue	10,000–28,000	Helium, some hydrogen
A	White	7,500–10,000	Strong hydrogen, some ionized metals
F	Light Yellow	6,000–7,500	Hydrogen and ionized metals, including calcium and iron
G	Dark Yellow	5,000–6,000	Metals and ionized metals, especially ionized calcium
K	Orange	2,500–3,500	Metals
M	Red		Titanium oxide and calcium

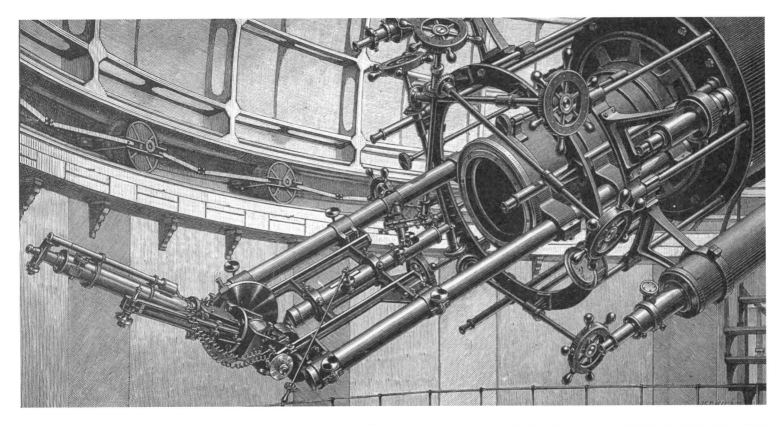

The Star-Spectroscope of the Lick Observatory, designed by James Keeler and constructed by John Brashear c.1898.

Cosmic Mnemonics

It can be difficult to remember the order of some important sequences in science, such as the colours in the visible spectrum, the planets in the solar system, or the classification scheme for stars. So, over the years mnemonics have been devised to help millions of students learn them. Mnemonics combine the first letter of the items in a sequence to be remembered via a memorable saying or rhyme in order to better recall them. For example, the order of the planets in the solar system might become *'Mary's Violet Eyes Made John Sit Up Nights (Pondering)'* for Mercury, Venus, Earth, Mars, Jupiter, Saturn, Uranus, Neptune (and Pluto – now a dwarf planet). The spectrum colours might become *Richard Of York Gave Battle In Vain* for Red, Orange, Yellow, Green, Blue, Indigo and Violet. For the stellar classification scheme we have *'Oh Be A Fine Girl Kiss Me'* to recall the sequence O, B, A, F, G, K, M! An extension of this sequence to cooler stars adds three more classes – R, N and S – resulting in the Harvard University mnemonic *'Oh Brutal And Foul Gorilla Kill My Roommate Next Saturday'*.

Pickering's Harem

While Edward Pickering was director of Harvard College Observatory, he received a substantial donation from the widow of a wealthy amateur astronomer to produce a star catalogue. The work involved a vast amount of complex mathematical calculations. He had been frustrated with the standard of work from his male assistants, claiming his 'Scottish maid', Williamina Fleming, could do better. This was no idle boast because she was a maths teacher who had fallen on hard times. He recruited her and many more women as 'female computers' to do the necessary calculations. They became known as 'Pickering's Harem' by members of the professional scientific community, from which women had previously been excluded. Many of the women, including Annie Jump Cannon, Henrietta Swan Leavitt, Antonia Maury and Fleming herself, became pioneering astronomers in their own right, helping to advance the sciences of astronomy and cosmology.

DOPPLER SHIFT

In addition to identifying the elemental composition of the stars and distant nebulae, the steady improvement in increasing the resolution of the spectra soon led to the realization that the positions of the various spectral lines did not precisely match the laboratory values. This was soon identified as evidence for a shift due to the speed of the light source: known as the *Doppler effect* (or Doppler shift). Austrian physicist Christian Doppler had described this phenomenon in 1842 in relation to sound. The Doppler effect accounts for the fact that the pitch of a moving source of sound, such as a police car siren, changes as it approaches and then recedes. As the car approaches, the wavelengths of sound are compressed, making the siren higher in pitch. As the car recedes, the wavelengths are stretched, making the siren lower in pitch. The Doppler effect was confirmed for light waves in 1848 by Hippolite Fizeau. In the case of light, the wavelengths are stretched as the light source moves away, shifting the spectral lines towards the longer wavelengths (red shift), and compressed as the light source approaches, shifting the spectral lines to the shorter wavelengths (blue shift).

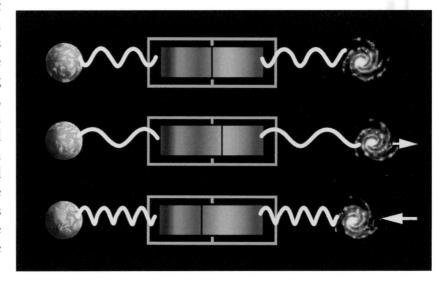

Spectral lines shifted by the Doppler effect.

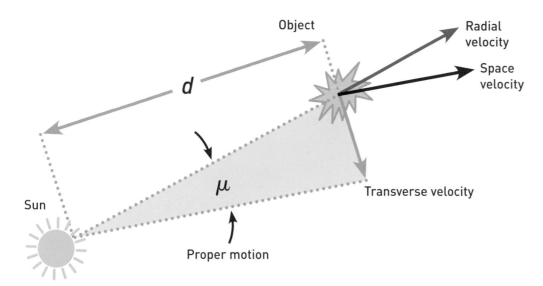

The actual motion of a star is a combination of its proper motion and its radial velocity.

SPACE VELOCITY ▶ *speed and direction in three-dimensional space.*

PROPER MOTION ▶ *speed and direction in two dimensions, as viewed from earth.*

RADIAL VELOCITY ▶ *speed away from (or towards) the Earth, as measured by Doppler shift.*

Space Velocity

The *space velocity* of an object is its movement (speed and direction) in three-dimensional space, and so when we look at stars in the sky and gauge their true speed through space, we have to specify their movement along three directions. Two of these directions are within the two-dimensional plane of the sky, as viewed from Earth. This motion is called the object's *proper motion* and can be given in degrees/year or, when the object's distance is known, in kilometres per second (km/s). The third dimension of motion is along the direction perpendicular to the plane of the sky, and this is called the object's radial velocity, which can be directly measured by the Doppler Shift without knowing the distance to the object. Combining the proper motion with the radial velocity gives the full space velocity of the object through space.

By 1887, photographic techniques had improved to the point that it was possible to measure a star's radial velocity from the amount of Doppler shift in its spectrum. By this means, the radial velocity of Aldebaran was estimated as 48 km/s (30 mp/s), using measurements carried out at Potsdam Observatory by Hermann C. Vogel. According to David Todd's 1897 book *A New Astronomy*, the Doppler effect by this time had resulted in the determination of the radial velocity for fewer than 100 stars, including Spica, Rigel, Aldebaran and Altair. The speeds were of the order of 20 to 50 km/s. The limit of accuracy was about 3 km/s. Even so, the speeds of over 80,000 km/h (50,000 mph) were considered to be incredible by any normal human standard at the time.

The Doppler Shift

Once it became possible to accurately measure the wavelengths of the spectral lines for starlight, their motion could be discerned from:

$$V = \frac{(\lambda - \lambda_0)}{\lambda_0} c$$

In this equation, V is the speed of the object emitting a specific spectral line at a wavelength of λ, for which the true wavelength of this line at rest on earth is given by λ_0. The quantity, c, is just the speed of light in the desired units such that c = 186,000 for V in miles/s and c = 300,000 for V in km/s. For example, the hydrogen Balmer-alpha line is seen in the laboratory at a wavelength of 656.45 nm, but observed in a distant galaxy at a wavelength of 756.45 nm. From the Doppler formula (756.45 - 656.45)/656.45 = +0.152, which means the galaxy is receding from the observer at 15.2 per cent the speed of light.

The Hertzsprung-Russel diagram related a star's temperature to its luminosity.

DISTANCES TO THE STARS

The chief problem faced by astronomers even by the late 1800s was the continuing presumption that stars were generally all of the same intrinsic brightness so that their faintness was a direct gauge of distance using the inverse-square law. This realization began to wane as the distances to more stars were determined using parallax methods, and from their distance and apparent brightness, stars with 0.1, 100 and even 1,000,000 times the solar light output were finally detected.

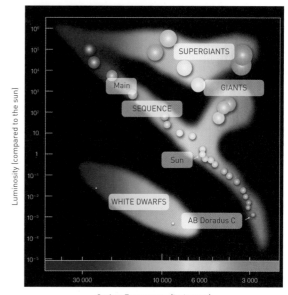

Surface Temperature (in degrees)

PARALLAX ▶ *the apparent shift in the position of an object in space when viewed from two different locations. Using geometry, the amount of this shift can be used to calculate the distance to the object.*

INVERSE-SQUARE LAW ▶ *that intensity (e.g. of brightness) is inversely proportional to the square of the distance from the source; that is, as distance increases intensity decreases using the formula $^1/_d{}^2$.*

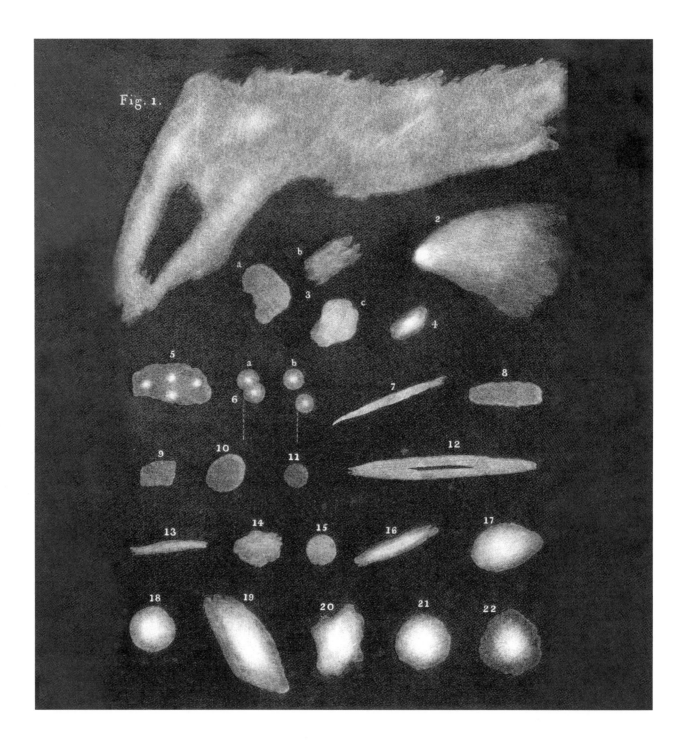

EXTRAGALACTIC NEBULAE

During the 18th century, astronomers such as the Englishman Edmund Halley and the Frenchman Nicolas de Lacaille made lists of a variety of odd-looking nebulosities observed through telescopes of steadily increasing size. The most famous of these early catalogues were those produced by the French comet hunter Charles Messier and the Anglo–German astronomer William Herschel. What were these nebulae, and what role did they play in cosmology?

Without any hard evidence, the German philosopher Immanuel Kant claimed, in his 1755 work *Immanuel Kant* *The Universal Natural History and Theories of the Heavens*, that these nebulae were really objects outside the boundaries of our Milky Way. This was a provocative idea at the time, especially since there was no indication that our galaxy, the Milky Way, had finite borders. In opposition to Kant, the French scholar Pierre-Simon Laplace proposed that nebulae were planetary systems in the making within the Milky Way, like the flat disk of our own solar system. Ultimately, Kant's view was largely supported by William Herschel in 1785, whose telescopic observations also led him to the idea of 'island universes' or separate galaxies, albeit within our own Milky Way galaxy. As Kant reflected in his book: '...*the feebleness of their light...demands a presupposed infinite distance: all this is in perfect harmony with the view that these elliptical figures are just (island) universes...*'

The nature of these nebulae became much clearer through a clever, but indirect, study begun by William Herschel's scientist son John Herschel. He plotted the locations of all known catalogued *John Herschel* nebulae on a graph showing the Milky Way along the equator of the sky, with the nebulae represented by dots. John Herschel wrote: '*The general conclusion which may be drawn from this survey, however, is that the nebulous system is distinct from the sidereal, though involving, and perhaps, to a certain extent, intermixed with the later.*'

Later, the American astronomer Cleveland Abbe used the Herschels' 1864 *General Catalogue of Nebulae and Clusters of Stars* with about 5,079 entries to show that the unresolved nebulae on the sky were outside the Milky Way while the resolved nebulae were mixed within the band of the stellar Milky Way. But the dispute over where nebulae were located remained an almost impossible one to resolve conclusively until actual distances could be obtained for them.

The technique of plotting new objects on an all-sky map was used by astronomers over 100 years later to discover where the mysterious objects called gamma ray bursts were located.

Opposite: Sketches by William Herschel of various nebulae seen with his telescope.

Annie Jump Cannon developed a system for the classification of stars.

SHAPLEY-CURTIS DEBATE

Stellar astronomy advanced dramatically with the advent of the study of the elemental composition of the stars and other cosmic matter using spectroscopy (see pages 27–8), photography and the construction of progressively larger telescopes. Nevertheless, even as late as 1920, there was still much that was not well understood. Even for stellar astronomy, the nature of the energy source that powered a star's luminosity was highly speculative, and the nature of the detailed structure of the Milky Way was still quite poorly known except for the immediate solar vicinity. Many nebulae had been spectroscopically studied and found to be rich in hydrogen and other gases. Large catalogues of stars had become commonplace, all grouped by spectroscopic type, according to the classification scheme developed by Annie Jump Cannon and Edward Pickering at Harvard University. There was, however, one category of objects that stubbornly resisted analysis and these were the extragalactic nebulae and their location in space.

The 'Great Debate' centred around whether nebulae were located within our Milky Way, the view championed by Harlow Shapley at the Mount Wilson Observatory, or truly extra-galactic, as favoured by Heber Curtis at the Lick Observatory. The debate was held at the National Academy of Sciences in Washington, DC in 1920. Each astronomer marshalled the best evidence for their position in a 45-minute session, during which over a dozen points were discussed. The debate largely resulted in a draw, as there was no clear consensus for either point of view. It would take another four years for this stalemate to be resolved.

In 1908, Henrietta Swann Leavitt, one of Edward Pickering's remarkable team of female computers at Harvard observatory, discovered a class of stars called Cepheid variables. These pulsated – expanded and brightened – in a strictly defined cycle. From studies of these stars in the Milky Way, Leavitt discovered that the time taken to complete one brightness cycle was directly related to their luminosity. If you know the luminosity of a star, you can calculate its distance using the inverse square law. This is much like seeing a lamplight in the distance and measuring its brightness in lux, but then being told that its luminosity is 100 lumens. From these two facts, brightness and luminosity, you can use the inverse-square law to accurately give the distance to the lamp.

American astronomer Edwin Hubble, using the powerful Hooker telescope at the Mount Wilson Observatory, went on to discover Cepheid variable stars in several spiral nebulae, including the Andromeda Nebula, making the announcement at the American Astronomical Society national convention of January 1924. In the Andromeda Nebula, their brightness clearly implied a vast distance of over 800,000 light years, placing this galaxy well beyond any estimate made by Shapley for the size of the Milky Way. The evidence developed by Hubble settled the Great Debate once and for all. The scale of the universe was indeed vast. In 1924, we had outgrown our own Milky Way and for the first time had an entire universe of strange new galaxies to explore and comprehend.

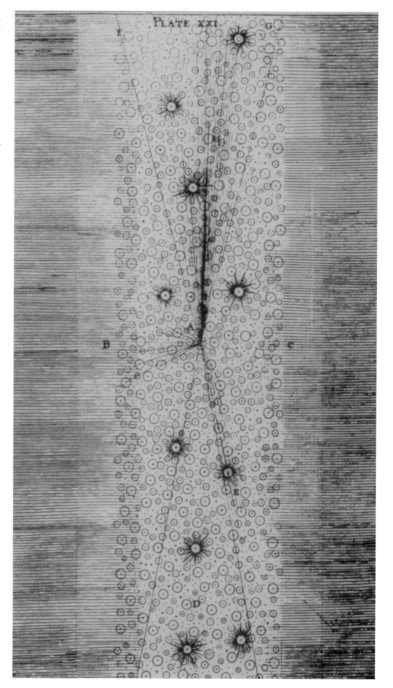

An early 19th-century idea for the shape of the Milky Way by Thomas Wright.

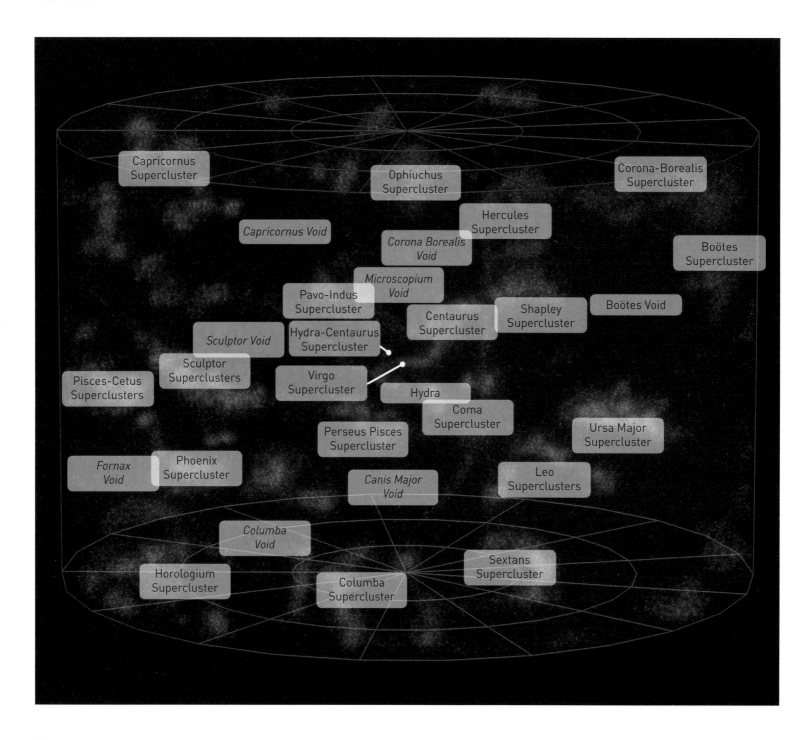

GALAXY CLUSTERS AND SUPERCLUSTERS

Galaxies were not randomly sprinkled across the sky but tended to group together into pairs and clusters. The largest of these were even known to 19th century astronomers, including the Virgo and Coma clouds. The first systematic catalogue of galaxy clusters was one created in the late 1940s by the American astronomer George Abell using the Palomar Observatory Sky Survey *George Abell* photographs commissioned by the National Geographic Society. Behind the confusing foreground clutter of the faint stars in the Milky Way, Abell found groupings of faint galaxies. Over 2,700 'Abell Clusters' were identified by 1958, and this catalogue was extended to include an additional 1,361 clusters in the southern hemisphere in 1989.

Abell's classification scheme for these clusters included the number of members and the degree of compaction of the cluster. The largest clusters of Richness Class 5 included the Coma and *Abell's Richness* Virgo clusters with over 1,000 members within a region of the sky no larger than that occupied *Classes* by the full Moon! Our Milky Way is one of 54 galaxies in the Local Group, but would barely make it into Abell's Richness Class 1. Amazingly, when the clusters were plotted on a map of the sky along with their diameters, Abell discerned a 'second-order' clustering into what we now call superclusters.

Superclusters are vast collections of clusters of galaxies spanning hundreds of millions of light years and in a variety of forms such as filaments and sheets. Although our Milky Way is a member of the Local Group, this group is on the outskirts of – and falling into – the Virgo Cluster. So our Milky Way is itself a member of the Virgo Supercluster containing more than 47,000 galaxies in *Superclusters* a volume some 110 million light years across. The largest identified supercluster is the Caelum Supercluster with 500,000 galaxies spanning just under one billion light years. However, the largest known structure in the cosmos, identified in 2013, is the Hercules-Corona Borealis Great Wall, which is nearly ten billion light years across and contains enough mass to form tens of millions of galaxies like our Milky Way.

Opposite: A perspective map of local superclusters of galaxies near our Local Group.

A sketch of the 'whirlpool' galaxy Messier 51 by the Earl of Ross.

PHOTOGRAPHY

Sir William Herschel used a 12-in (30-cm) telescope between 1773 and 1800 to survey Charles Messier's catalogue of comet-like objects, sketching what he saw in detail. The Third Earl of Rosse used a much larger 72-in (183-cm) telescope, the largest of its kind from 1845 to 1917, and also created many sketches of what he saw at the eyepiece. These sketches remained the illustrative currency of astronomy books as late as Charles Young's *A Text-book of General Astronomy* published in 1889, and his subsequent book *The Elements of Astronomy* published in 1892. What was needed was a better true-to-life means for capturing an object's actual appearance without the need for human-made sketches. The advent of photography in the 1800s was the answer.

Although there were a number of silver-based photographic techniques being invented before 1838, the method employed by Louis Daguerre became the most popular by the 1840s. The technique was applied for the first time to astronomical photography in 1840 by John William Draper, who 'daguerrotyped' the full Moon, and then in 1845 by Louis Fizeau, who took the first detailed photo of the Sun showing sunspots. By the 1850s, it became possible to photograph stars and, in 1880, physician and amateur astronomer Henry Draper used a 50-minute exposure to capture the likeness of the Great Nebula in Orion. From that time on, it was simply a matter of making photographic technology more compact, faster, and more efficient before it was regularly used in conjunction with telescopes of ever-increasing size to study planets as well as faint astronomical objects.

David Todd at Amherst College published *A New Astronomy* in 1897, which featured dramatic astronomical photographs by George Ellery Hale, Isaac Roberts and Edward Barnard. The *Textbook of Astronomy* by George Comstock at the University of Wisconsin in 1901 followed suit with many additional photographs by George Ellery Hale, James Keeler and the Paris Observatory. Among the first popularizations of astronomy to feature photographs was Garrett Serviss' *Curiosities of the Sky*, published in 1909, with its striking full-page images provided by James Keeler at the Lick Observatory.

Spectroscopy was also in need of larger telescopes to gather more light for making better spectra of a variety of faint stars and nebulae. In 1872, Henry Draper made the first photographic image of the spectrum of the star Vega, and this method was perfected so that by 1885, Edward Pickering at the Harvard College Observatory was able to use this recording technique in an ambitious program of stellar spectral classification with tens of thousands of spectra recorded on photographic plates. The race was now on to build larger telescopes to study a variety of galactic and extragalactic objects in increasingly higher detail.

NAMES TO KNOW: PHOTOGRAPHING THE UNIVERSE

John William Draper (1811–1882)

Louis Fizeau (1819–1896)

Henry Draper (1837–1882)

Edward Pickering (1846–1919)

David Todd (1855–1939)

Edward Pickering

Giant eyes on the skies

Beginning in the early 20th century, a new technique was developed for depositing a silver surface on glass. This led to the construction of large research-grade reflecting telescopes with glass mirrors housed on remote mountain tops away from city lights and above much of the obscuring atmosphere. The 152-cm (60-in) Hale telescope in 1908 was followed by the 254-cm (100-in) Hooker telescope in 1917 on Mount Wilson outside Los Angeles. In 1948, the 508-cm (200-in) Hale reflector was constructed on Mount Palomar. By the 1990s, telescopes were being built at even higher and more remote locations such as Mauna Kea in Hawaii and on the 4267-m (14,000-ft) peaks of the Andes Mountains in Chile. Modern telescopes used for cosmological research include such mammoths as the Very Large Telescope 820-cm (323-in) in Chile, the Keck telescopes 1,000-cm (394-in) on Mauna Kea, and the *Gran Telescopio Canarias* 1039-cm (409-in) in the Canary Islands. Under development are even larger, ground-based telescopes such as the Extremely Large Telescope (ELT) in Chile, whose mirror will be 40 m (130-ft) in diameter, and the Large Synoptic Survey Telescope with a diameter of 8 m (27 ft). When equipped with electronic cameras and sensitive spectroscopic equipment, they can detect and study the light from galaxies over ten billion light years distant.

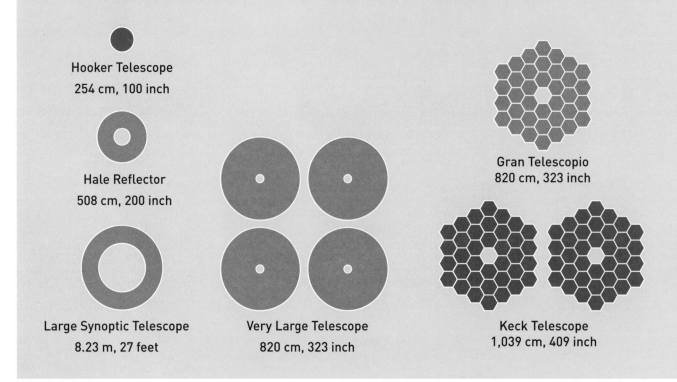

Hooker Telescope
254 cm, 100 inch

Hale Reflector
508 cm, 200 inch

Large Synoptic Telescope
8.23 m, 27 feet

Very Large Telescope
820 cm, 323 inch

Gran Telescopio
820 cm, 323 inch

Keck Telescope
1,039 cm, 409 inch

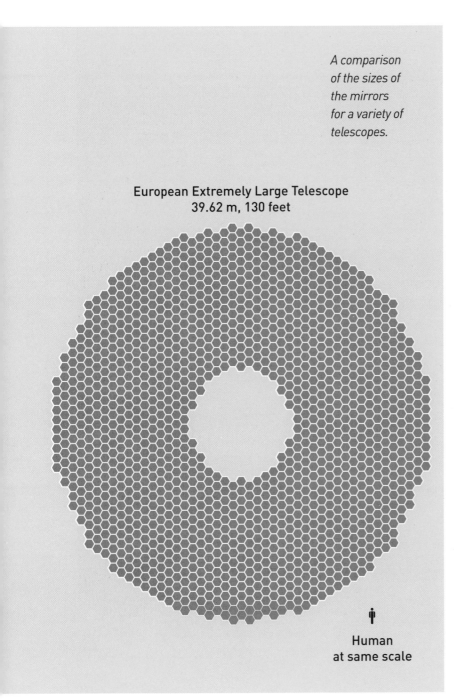

A comparison of the sizes of the mirrors for a variety of telescopes.

European Extremely Large Telescope
39.62 m, 130 feet

Human
at same scale

Standard film-based black-and-white astrophotography held sway until the mid-1930s, when the first colour films were introduced and there were sporadic attempts by observatories to create photographs of celestial objects. After a new generation of films was introduced in the 1950s, astronomers including William Miller at the Mount Wilson and Palomar Observatories began to use the 122-cm (48-in) Schmidt telescope to take long-exposure colour photos of many popular objects, including famous nebulae such as M 42 in Orion and galaxies such as M 32 in Andromeda.

No matter how well film or glass-plate photography worked, it was still very cumbersome to carry photographic plates to the telescope, make the exposure, and then bring the plates back to the darkroom to process them using chemical baths. The slightest flaw in the developing process could destroy hours of time at the telescope to gather the faint images. There were many improvements in photographic technology, which accelerated during the first half of the 20th century in the quest for faster speeds, shorter exposure times, and simpler developing techniques, leading to the colour astrophotography of the second half of the century. A major

A modern photograph of the Orion Nebula with the VISTA telescope in Chile.

stimulus to advancing this technology came from military applications and from the fledgling NASA space programme. By the 21st century, purely electronic means would often be used to capture images.

1965

In 1965, NASA's Mariner 4 spacecraft flew by Mars and captured a few dozen images of its cratered landscape. It used a scanning video tube whose analogue light intensity output was 'digitized' into a string of numbers and telemetered back to Earth for reconstruction into an image.

1975

Then, in 1975, Eastman Kodak engineer Steven Sasson adapted the new charge-coupled device (CCD) solid state technology developed by Fairchild Semiconductor Electronics in 1973, to record the first true digital image. The 'CCD' array of 10,000 'pixels' took 23 seconds to capture its first image, and was just an engineering test concept.

1988

But in 1988, the first commercial digital camera was the Fuji DS-1P, which, unfortunately, was expensive ($5,000) and not commercially popular. Astronomers soon took advantage of this new experimental imaging technology and its amazing prospects. CCD-based images were easy to use and manipulate using computers, and their light sensitivity was far more uniform across the visible spectrum than photographic film emulsions. They could even be tailored to be sensitive to infrared light.

In 1976, James Janesick and Gerald Smith at the NASA Jet Propulsion Laboratory and the University of Arizona, obtained images of Jupiter, Saturn and Uranus using a CCD detector attached to the 155-cm (61-in) telescope on Mount Bigelow in Arizona. By 1979, the Kitt Peak National Observatory had mounted a 320×512 pixel digital camera on their 1-m (40-in) telescope and quickly demonstrated the superiority of CCDs over photographic plates. Since the 1990s, there has been huge pressure to fill the entire focal plane of large telescopes with 'mosaicked' digital arrays numbering in the hundreds of millions of pixels. In one exposure, the astronomer could now capture the light from huge areas of the sky all at once, rather than in hundreds or thousands of time-consuming, individual film exposures.

In addition to greatly increased telescope sizes and camera array formats, a technique known as adaptive optics can manipulate the surfaces of mirrors thousands of times a second, and with the help of laser 'guide stars' almost completely eliminate the effects of atmospheric twinkling. This causes star images to be nearly as sharp as those available from space-based telescopes above the atmosphere, with dramatic improvements in clarity for far less cost.

CCD detector

Guide stars

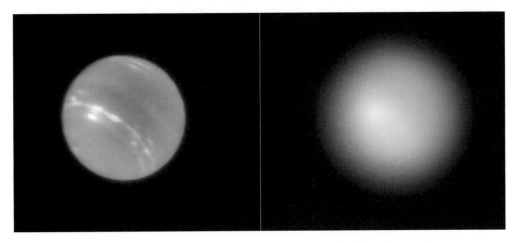

Adaptive optics　　　　No Adaptive optics

Opposite: An artist's impression of one of the largest modern telescopes under construction, the Extremely Large Telescope in Chile.

WOMEN ASTRONOMERS

Annie Jump Cannon (1863–1941) was an American astronomer whose cataloguing work was instrumental in the development of contemporary stellar classifications. Edward Pickering hired her at the Harvard College Observatory as an assistant in 1896. She is credited with the creation of the Harvard Classification Scheme based on the strength of the *Balmer absorption lines* seen in the stellar spectra, which was the first serious attempt to organize and classify stars based on their temperatures and spectral types.

Her effort involved classifying the spectra of over 350,000 stars. Astronomers were previously using a 22-type alphabetic scheme, but after finding duplicates this was reduced to a smaller series and re-arranged by Jump Cannon to reflect a temperature order, giving us the modern O, B, A, F, G, K, M system. She was nearly deaf throughout her career due to scarlet fever in 1893. A suffragist and a member of the National Women's Party, she was admitted into the Royal Astronomical Society in 1914, and was the first woman to be given an honorary doctorate at Oxford in 1925.

Cecelia Payne (1900–79) won a scholarship in 1919 to Newnham College, Cambridge University, where she read botany, physics, and chemistry. Here, she attended a lecture by the English astronomer, mathematician and physicist Arthur Eddington on his 1919 expedition to photograph the stars near a solar eclipse as a test of Einstein's general theory of relativity (see page 67). This sparked her interest in astronomy. At Harvard under Harlow Shapley, she wrote her 1925 doctor's dissertation on 'Stellar Atmospheres'. She used Jump Cannon's stellar classification scheme, along with the newly developed theory of *quantum mechanics*, to prove that stars were made mostly of hydrogen and helium, and that the spectra seen were a complicated combination of temperature, ionization state and abundance factors. Astronomer Otto Struve called it 'undoubtedly the most brilliant PhD thesis ever written in astronomy'.

Cecilia Payne was a leading British astronomer who was one of the first to apply quantum mechanics to the field.

Chapter Three
THE RELATIVITY REVOLUTION: SPACE 2.0

Dimensionality – Cartesian Co-ordinates – Newtonian Physics and Universal Gravitation – Newton and the Size of the Universe – The Dark Sky Mystery – Relativity – Spacetime – Gravity – Curved Spaces and Higher Dimensions – Gravity Lenses – Proving Einstein was right – Gravity Waves

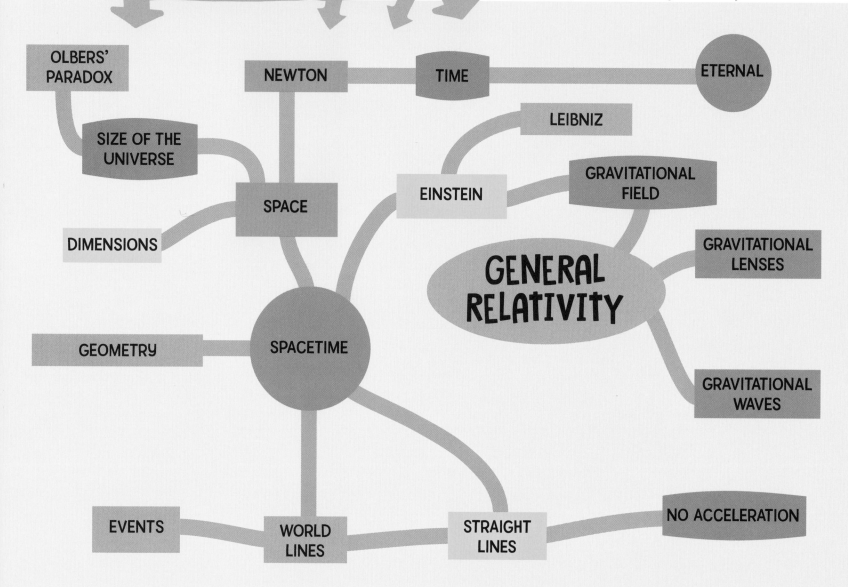

OLBERS' PARADOX

NEWTON

TIME

ETERNAL

LEIBNIZ

SIZE OF THE UNIVERSE

EINSTEIN

GRAVITATIONAL FIELD

SPACE

GRAVITATIONAL LENSES

DIMENSIONS

GENERAL RELATIVITY

GEOMETRY

SPACETIME

GRAVITATIONAL WAVES

EVENTS

WORLD LINES

STRAIGHT LINES

NO ACCELERATION

DIMENSIONALITY

The idea that there are three dimensions to space is at least as old as Euclid's geometry. But the prevailing view was that there were no more than three. Simplicius of Cilicia noted in 600 CE that 'the admirable Ptolemy in his notebook "On Distances" well proved that there are not more than three distances'. Stifel, in the 16th century, described '...*going beyond the cube just as if there were more than three dimensions...which is against nature*'. John Wallis, a contemporary of Isaac Newton, protested that '...*Length, Breadth and Thickness take up the whole of Space. Nor can Fansie imagine how there should be a Fourth Local Dimension beyond these Three*'. But this didn't stop enterprising mathematicians in the 1800s from considering other possibilities beyond Euclid's restrictive plane geometry, literally with the stroke of a pen.

Every point in 3-D space is defined by exactly three numbers defining a co-ordinate system.

CARTESIAN CO-ORDINATES

Invented by the French philosopher and mathematician René Descartes, *cartesian co-ordinates* revolutionized mathematics in the 17th century by creating the first systematic link between Euclidean geometry and algebra. Cartesian equations can be used to describe geometric shapes such as curves. For example, a circle of radius 2 can be described as the set of all points whose co-ordinates satisfy the equation $x^2 + y^2 = 4$. Therefore, the location of a single point on a flat surface such as a sheet of paper can be described by just two numbers, given with reference to an 'x axis' and a 'y axis'. This number 'two' is called the dimensionality of the space, or more colloquially we say the space is two-dimensional. Similarly, there are only three numbers needed to describe the location of any point in three-dimensional space, which introduces a third 'z axis'. This idea of dimension is completely general, and doesn't have to refer to the properties of physical space. It can also include other properties such as time – the 'fourth dimension'.

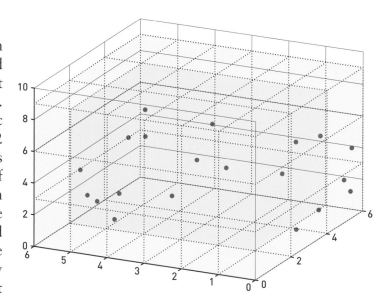

The geometry of a four-dimensional world led to a number of very interesting properties for three-dimensional objects, the study of which is known as *topology*. For example, the American astronomer Simon Newcomb demonstrated that you could turn a closed shell inside out without tearing. Mathematician Felix Klein showed that knots could not be tied in the fourth dimension. Giuseppe Veronese at Padua University proved that a body could be removed from a closed box without breaking its walls.

TOPOLOGY ▶ *the study of the properties of multi-dimensional space which are conserved under continuous deformations such as stretching, twisting, crumpling and bending but not tearing or gluing.*

Flatland

Victorian schoolmaster Edwin Abbott Abbott tried to put across the concept of a fourth dimension in his book *Flatland: A Romance of Many Dimensions* (1884). To simplify the issue, he described how a three-dimensional 'alien' would be viewed by the beings living on a two-dimensional world. As the 3-D alien passed through the 2-D world it would appear first as a dot, rapidly expanding in circular area and then shrinking and disappearing again. A 2-D being would have to try to understand the nature of this extra dimension using 2-D concepts. If a 2-D being encountered a bump on the sphere it would be aware of a force (gravity) acting upon it, slowing its movements as it went up one side and speeding it up as it came down the other side. But it would be hard pressed to explain this phenomenon except through mathematics.

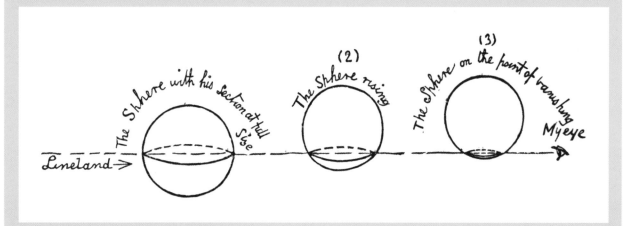

'"See now, I will rise; and the effect upon your eye will be that my Circle will become smaller and smaller till it dwindles to a point and finally vanishes." There was no "rising" that I could see; but he diminished and finally vanished.'
From Flatland: A Romance of Many Dimensions (1884), *by Edwin Abbott Abbott.*

NEWTONIAN PHYSICS AND UNIVERSAL GRAVITATION

Sir Isaac Newton revolutionized how physicists thought about motion and gravity by formulating a detailed theory of how bodies move. But hidden behind the scenes of this 'New Physics' was his idea of what three-dimensional space and time were like. Because all of his equations involved the position of bodies in space defined by their spatial co-ordinates (x, y, z) and the coordinate of time (t), he took the position that there was an Absolute Cosmic Reference Frame through which bodies moved. According to Newton, '…*God is everywhere present, and by existing everywhere and always, he constitutes duration and space*'. In other words, space is an attribute or extension of God, forming the Absolute reference frame for existence. But the philosopher Bishop Berkeley considered Newton's idea of an absolute physical space to be empty of meaning because, to his way of thinking, a vacuum stripped of all physical objects was also stripped of its geometric content.

Absolutists v Relativists

Berkeley was among the first 'relativists' who considered that motion was meaningful only when measured relative to another body – an idea later developed by the German mathematician Gottfried Leibniz. He and Leibniz also anticipated by 200 years Albert Einstein's view that the geometric properties of space are predicated upon the existence of the matter that fills space. Absolute Space, as something aloof from matter and pre-existent, was, for Berkeley, Leibniz and later Einstein, a complete absurdity, and even philosophically offensive.

Meanwhile, Newton, if he considered at all the issue of why gravity occurs, never wrote about it at great length. In his magnum opus the *Principia* he wrote '… *But hitherto, I have not been able to discover the cause of those properties of gravity from phenomena and I frame no hypothesis…*' Yet in an informal letter to Henry Oldenburg in 1675, he suggested: '*It is to be supposed therein, that there is an aetheral medium much of the same constitution with air, but for rarer, subtler and more strongly elastic. For the electric and magnetic effluvia, and the gravitating principle, seem to argue such variety. Perhaps the whole frame of nature may be nothing but various contextures of some certain aetheral spirits or vapours, condensed as it were by precipitation… Thus perhaps may all things be originated from aether…*"

Gottfried Wilhelm Leibniz

Leibniz was a prominent German polymath, mathematician and philosopher of the 18th century whose optimism about the world is famously displayed in his notion that we live in the best of all possible universes that could have been created by God. As a mathematician, he co-invented calculus though disputed heavily by his contemporary Isaac Newton. Perhaps more notably for physics, his concept of relativism would become the foundation for 20th-century Relativity Theory, in which the Newtonian framework of a fixed space and time is replaced by objects themselves creating space and time via their inter-relationships.

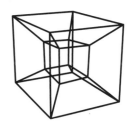

A 3-D representation (projection) of a 4-D cube, sometimes called a tesseract.

NEWTON AND THE SIZE OF THE UNIVERSE

Newton believed that the universe had to be infinite because of a rather compelling argument based upon his gravitational physics. If each body was acted upon by every other body in the universe, then in order for there to be no movement of the stars in the sky (presumed to define the limits of the cosmos) the universe would have to be immense or even infinite. If this were not the case, in a finite universe the entire ensemble of material bodies would collapse under their own mutual gravitational attractions. Only in an infinite universe could there exist the opportunity for the infinite multitudes of bodies to add their gravitational attractions upon each other so that the net forces upon any one of them would cancel. The fact that no one sees the stars swirling around the sky at night from decade to decade meant to Newton that the universe was infinite in spatial extent and filled more or less with a uniform population of bodies everywhere. Newton also wrote that an *'infinite and eternal'* divine power coexists with space, which *'extends infinitely in all directions'* and *'is eternal in duration'*. As Newton noted in his 1666–68 unpublished manuscript *De gravitatione… 'if the starry heavens were of finite extent they would "fall down to the middle" and there "compose one great spherical mass".'*

Infinite universe

Olbers' Paradox can be simulated in a dense forest where all sightlines end on the trunk of a distant tree.

THE DARK SKY MYSTERY

Kepler's mathematics could circumscribe the dimensions of the planetary region, the stellar sphere remained unconstrained and, as some had previously asserted, infinite. Kepler argued that the universe had to be finite, otherwise the night sky could never be completely dark. The English astronomer Thomas Digges had already considered this 'dark sky' problem decades before Kepler. He pioneered the idea of an unlimited universe 'filled with the mingling rays of countless stars'. But this led to the problem of why the night sky was dark, and not filled with the light of 'countless stars'. This idea was independently resurrected in 1826 by the German astronomer Heinrich Wilhelm Olbers and is now called 'Olbers' Paradox'.

Olbers' Paradox

Imagine you are standing in a deep forest. No matter in which direction you look, your view to the distant horizon will encounter the trunk of a tree. In an infinite universe, every line of sight will eventually land on the surface of some star, somewhere in the depths of space. Even though light intensity decreases with the square of the distance, and distant stars should contribute less and less to the light of the sky, at every distance in a uniform universe the number of stars at that distance will increase with the square of the distance. The two effects of light dimming and population increasing cancel out at every distance, adding to a constantly increasing lighting of the sky. In such a universe, the entire sky would glow with the brightness of the surface of a star! Olbers' Paradox seemingly could only be resolved if the Universe were either finite in space or limited in time.

In 1848, the author Edgar Allan Poe offered another solution to Olbers' Paradox in 'Eureka: A Prose Poem'. *'Were the succession of stars endless, then the background of the sky would present us a uniform luminosity, like that displayed by the Galaxy – since there could be absolutely no point, in all that background, at which would not exist a star. The only mode, therefore, in which, under such a state of affairs, we could comprehend the voids which our telescopes find in innumerable directions, would be by supposing that the distance [of] the invisible background [is] so immense that no ray from it has yet been able to reach us at all.'* Space was vast but not necessarily infinite, and it took a long time for light to reach us from the stars that would fill in the sky at the present moment.

RELATIVITY

In the 1800s, investigations of electric charges and magnetism had turned up a number of important laws, leading to a new theory of electrodynamics created by James Clerk Maxwell. This theory tied together these separate findings into a unified mathematical theory that described how charged particles and their movement in currents led to magnetism, and even electromagnetic radiation discovered by the German physicist Heinrich Hertz and identified with light itself.

James Clerk Maxwell

ELECTRODYNAMICS ▶ *how the movement of charged particles in currents leads to magnetism.*

Electrodynamics

Experiments with charged particles in 1785 by Charles-Augustin de Coulomb led to the discovery and mathematical description of electrical currents by André-Marie Ampère in the early 1800s, and the production of magnetic fields from these currents by Hans Christian Ørsted in 1820. This new property of moving charges was called electromagnetism and led to the invention of the electric motor and the generator. In 1831, Michael Faraday discovered that changing currents in one wire could induce currents in a neighbouring wire: a process called electromagnetic induction, leading to the invention of the transformer. In the 1860s, James Clerk Maxwell was able to mathematically describe all of the experimental phenomena related to charged particles, currents and magnetism known by that time in a set of four equations, which are known as Maxwell's Equations of Electrodynamics. From these equations, a mathematical 'wave equation' could be derived, which represented an electromagnetic wave, later identified with radio and light emission in their various forms.

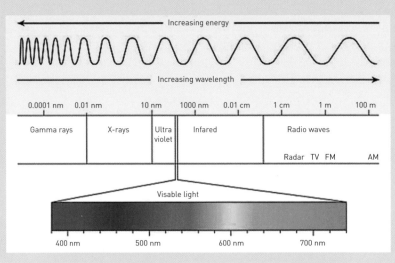

In 1864, Maxwell tackled the problem of how to make his theory of electrodynamics work so that it would describe charged, moving bodies, but his efforts soon uncovered a glaring inconsistency; his equations did not look the same if they were described from the vantage point of some other reference frame moving at a constant speed. This is a difficult and subtle problem to understand. Suppose you and your friend were performing an experiment on the induction of currents in a wire by a moving magnet. Suppose, also, that your friend was doing his experiment from the back seat of a car moving away from you. Even though you might both be performing identical experiments, you will observe your moving friend's electric and magnetic fields to be different from the ones you see in your own experiment. How could identical experiments appear different just because one of you was moving? For a time it was thought that the laws dictating the behaviour of Maxwell's electromagnetic waves in Marconi's transcontinental radio messages really were separate from those of Newtonian mechanics, which described planets and cannon balls. It wasn't until 1905, when the German physicist Albert Einstein announced his Special Theory of Relativity, that a consistent interpretation emerged.

Special Relativity provided four new phenomena for physicists to reckon with. As you approach the speed of light, time slows down, length gets shorter and mass increases. The most popular phenomenon, and by far the most dramatic prediction, is given by $E = mc^2$, which states that energy (E) and mass (m) are interchangeable physical concepts in Nature. If you could convert one gram of matter into energy, you would release more energy than is present in most atomic bombs. The Sun must convert nearly two million tons of matter into pure energy each second, in order to shine as brightly as it does. For the induction experiment, the situation is different from what one intuitively imagines.

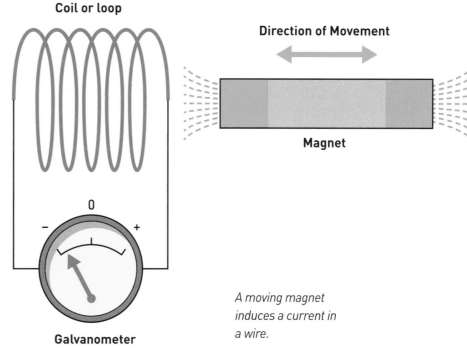

A moving magnet induces a current in a wire.

The experiment you are conducting in your reference frame is at rest with respect to you, but your partner in the car is moving away from you at a particular velocity. Relative to the car passenger, his currents are also travelling at the same speed as your currents, so both you and the car passenger will measure the same currents and the same magnetic fields. But if you look at the car passenger's currents, they are travelling at a different speed from yours due to the relative motions between the reference frames. That leads you to see a different intensity of magnetic fields in the car experiment. When the equations of special relativity are applied, with the finite speed of light (and currents) included, the car experiment measurements can now be exactly calculated in terms of the 'proper' currents and fields you measure in your experiment – and the mystery of Maxwell's Paradox is resolved.

SPACETIME

The constancy of the speed of light in all non-accelerating reference frames means that when you translate your measures of space and time intervals from one frame to the next the new co-ordinates become a mixture of space and time units! The term 'spacetime' was coined by the German mathematician Hermann Minkowski to express this intertwining of space and time in all descriptions of the physical world. *'The views of space and time which I wish to lay before you have sprung from the soil of experimental physics, and therein lies their strength. They are radical. Henceforth space by itself, and time by itself, are doomed to fade away into mere shadows, and only a kind of union of the two will preserve an independent reality.'* Minkowski was the first to treat the histories of particles in time and space as 'worldlines' in a four-dimensional flat space.

This new arena included three dimensions of space and one dimension of time: the four-dimensional, spacetime continuum. The idea of spacetime is an incredibly useful physical idea, but it led to a major philosophical problem. The true arena for physics was a four-dimensional spacetime 'block' in which worldlines traced out the complete histories of every particle, literally from birth to death.

Herman Minkowski

Minkowski was a German mathematician who specialized in the properties of N-dimensional geometry after the discoveries by Berhard Riemann in the mid-1800s. As a former teacher of Albert Einstein at the Eidgenössische Polytechnikum in Zurich, he investigated Einstein's recent Special Theory of Relativity in 1907. He discovered that its particular space and time co-ordinate transformation equations could be treated as the geometric properties of a 4-dimensional 'spacetime' continuum. Minkowski went on to develop all of the nomenclature we use in relativity including events, worldlines and the term 'spacetime' itself. His geometry of special relativity required a flat Euclidean-like space with no curvature, which is called Minkowski Spacetime.

A 2-D version of Minkowski's 4-D spacetime with time as the fourth dimension.

WORLDLINE ▶ *the path that an object takes in 4-D spacetime; it includes the object's position at different moments in time from the past to the future as well as its location within 3-D space.*

BLOCK UNIVERSE ▶ *the theory that all objects and events in the universe – past, present and future – are all together in one 4-D block.*

But this necessarily meant that it was a completely timeless view of nature. It eliminated the significance of the present moment we call 'now', and replaced it with a perspective in which nothing actually moved. From the spacetime viewpoint, particles do not move, they are simply viewed all-at-once from their complete historical perspective. Through a simple mathematical and logical argument, all of the 'nows' in the block universe were equally real and so past, present and future were equally real as well, and in some sense, they are continually in existence. Nevertheless, this new and mind-numbing perspective allowed special relativity to be ruthlessly practical, predicting new and testable phenomena in nature and resolving the paradoxes implied by Maxwell's theory.

Every event in spacetime is defined by a 'light-cone' defining past, present and future.

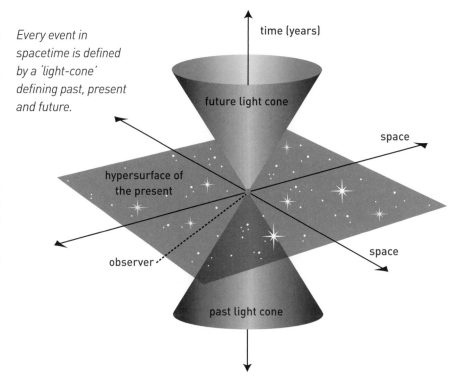

GRAVITY

Despite its tremendous initial successes, there was a glaring deficiency in Special Relativity: there was no way to describe the influence of gravity. Special Relativity is a theory in which the relative speeds between reference frames are constant so all worldlines are Euclidean 'straight lines'. Since gravity introduces accelerations between widely separated reference frames, Special Relativity is only valid over brief intervals of time when motions seem constant. Also, since gravity varies from point to point in space, Special Relativity is only applicable in small regions of space.

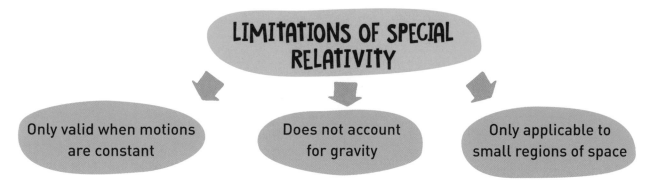

LIMITATIONS OF SPECIAL RELATIVITY

Only valid when motions are constant

Does not account for gravity

Only applicable to small regions of space

Between 1905 and 1912, Einstein worked on many aspects of how to go beyond Special Relativity before he finally published his *Theory of General Relativity* in 1915. Einstein knew that the theory of gravity would have to be very complicated, but simplify to the familiar physics of Newtonian gravitation when motions were much slower than light and gravity was weak. Sometime between 10 August and 16 August 1912, he began to realize that an entirely new geometry for spacetime was needed. The flat Euclidean geometry used by Special Relativity did not fit the bill because all of the reference frames would be related by the simple transformation that did not include gravity.

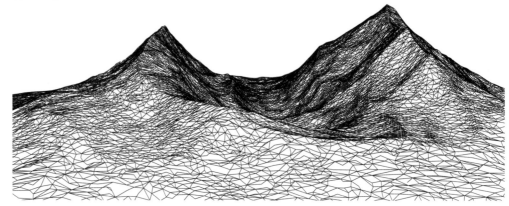

Even the bumpy surface of Earth can be tiled by 2-D flat pieces.

Imagine tiling a sphere with flat postage stamps. If you make the stamps small enough you can reproduce the shape of the sphere very closely. On the surface of each stamp there is no curvature at all. With Special Relativity, each of these stamps represents the locally flat Minkowski spacetime. The true spacetime, however, like the surface of the sphere, is very curved with gravitational distortions causing acceleration changes between the 'stamps'.

Einstein tried to work out how to describe the overall acceleration from the vantage points of small pieces of spacetime, but this proved to be a very hard problem to explore because he lacked the mathematical tools he needed. Einstein hadn't realized that mathematicians during the 19th century had long since gone far beyond the traditional flat geometry of Euclid that Einstein was taught. They had created a completely new kind of geometry with its own special rules that applied to curved spaces of an arbitrary number of dimensions. He heard about this new mathematics of curved geometry from his friend the mathematician Marcel Grossman. He quickly recognized that the mathematics of curved space was exactly what he was looking for. Einstein saw that acceleration would appear as the curving of worldlines in spacetime, but he needed some way to relate this curvature to the properties of matter itself. The way this was developed in detail is key to understanding the nature of space and time in modern cosmology.

Beyond Euclidean geometry

CURVED SPACES AND HIGHER DIMENSIONS

Surveyors have to find a way of measuring small portions of the surface of Earth under conditions where its two-dimensional surface is warped by mountains and other irregularities.

In 1827, Carl Friedrich Gauss developed the idea that the geometric properties of a two-dimensional surface embedded in a three-dimensional Euclidean space can be explored by studying the equations representing the distance or metric interval between two points within the surface. The Pythagorean Theorem is one example of such a metric interval, which states that the distance (dS) between any two points in three-dimensional space is given by:

$$dS^2 = dx^2 + dy^2 + dz^2$$

This can be expressed in mathematical shorthand for different kinds of curved surfaces as:

$$dS^2 = g_{ij}\, dx^i\, dy^j$$

The quantity g_{ij} is called the fundamental metric tensor, which encodes mathematical information about the surface geometry.

Carl Friedrich Gauss (1777–1885) pioneered the mathematics for three-dimensional surfaces.

Carl Friedrich Gauss

Gauss is often called one of the most brilliant mathematicians in history, and his work during the early 19th century set the stage for advancements in mathematics and physics. In his surveying work for the Kingdom of Hanover in 1818, he developed the basic mathematics of curved three-dimensional surfaces, which later led to the N-dimensional geometry discoveries by Bernhard Riemann in the mid-1800s and the tensor mathematics of curvature needed to complete Einstein's General Theory of Relativity in the early-1900s. Gauss discovered that the geometry of space, in particular its curvature and topology, can be deduced by making measurements from within the space via a careful analysis of 'deficit angles' in surveying. Gauss also made many contributions to the study of magnetism, helping to develop its foundations as a mathematical theory.

The next step came with the extension of Gauss's approach to surface geometry so that it worked in any number of dimensions, not just two. This was accomplished by the German mathematician Georg Riemann in 1854. He defined a mathematical measure of a surface, in particular its curvature from point to point, based solely on the properties within the surface as expressed by Gauss's metric formula.

$$R_{mnij} = \frac{1}{2}\left[\frac{\partial^2 g_{jm}}{\partial x^i \partial x^n} - \frac{\partial^2 g_{nj}}{\partial x^i \partial x^m} - \frac{\partial^2 g_{im}}{\partial x^j \partial x^n} + \frac{\partial^2 g_{ni}}{\partial x^j \partial x^m}\right]$$

Riemann's curvature formula

Suppose in a three-dimensional Euclidean space you used Cartesian co-ordinates to define the positions of points x, y, z. Riemann's curvature formula would tell you that in this choice of a co-ordinate system, $R_{mnij} = 0$ and so space is flat. If you now used a new co-ordinate system to measure the same points – 'spherical co-ordinates', for example – your points would be labelled according to two angles and a radial distance from the centre of your co-ordinate frame $(r, \theta\phi)$. If you again computed Riemann's curvature formula it would once again tell you that space is flat! Riemann's curvature formula tells you about whether your three-dimensional space is curved or not by measuring the properties of space from inside it. Also, once you know g_{ij}, you can use Riemann's mathematical tools to calculate the shortest distance between any two points along a path called a geodesic curve.

Georg Bernhard Riemann created a formula for determining the curvature of a surface.

GEODESIC CURVE (OR ARC) ▶ *the shortest distance between any two points on a curved surface. In Flatland (see page 52), a 2-D being living on a sphere and following a geodesic curve would think it had travelled in a straight line, although it had actually described an arc.*

It doesn't matter what co-ordinate system you happen to choose to describe where the points are located. Einstein saw that if you replaced the mathematical 'co-ordinate system' idea with the physical 'reference frame', and 'geodesic curves' with worldlines, he had a theory that included acceleration as the curvature of worldlines where the worldlines were the geodesic curves linking points in spacetime. These geodesic curves would be the paths followed by matter or by light rays.

Georg Friedrich Bernhard Riemann
Often called Bernhard Riemann, he was a German mathematician and former student of Carl Friedrich Gauss at the University of Gottingen in 1847. In 1853, Gauss asked Riemann to prepare a detailed proposal for his doctoral dissertation on geometry. From this, Riemann developed the foundations of N-dimensional non-Euclidean geometry using the geometric tools provided by Gauss together with advances in the field of differential geometry. His doctoral dissertation *'On the hypotheses which lie at the bases of geometry'* was not published until after his death from tuberculosis in 1866, but established an entire school of Riemannian Geometry, setting the stage for Einstein to develop his General Theory of Relativity some 40 years later.

Airliners follow the shortest routes (geodesic curves) on the surface of the Earth for long-distance flights.

Einstein used the language of Riemannian geometry, together with the tensor mathematics developed by the Italian mathematician Gregorio Ricci-Curbastro, to express the relationship between the geometry of spacetime and gravity – Einstein's now famous field equation for gravity:

$$R_{\mu\nu} - {}^{1}/{}_{2}Rg_{\mu\nu} = -\frac{8\pi G}{c^4}\,T_{\mu\nu}$$

The first thing you notice is that this beautiful but mysterious-looking formula bears no resemblance to Newton's Law of Gravity, which is written as:

$$F = \frac{GMm}{r^2}$$

In fact, you can recover Newton's equation for gravity from Einstein's relativistic equation for gravity in the case where the gravitational field (curvature) is weak and the speeds of particles are very slow compared to light speed. Einstein's equation, $T_{\mu\nu}$, called the energy–momentum tensor, tells how matter and energy are distributed at all points in spacetime. The Ricci curvature tensor, $R_{\mu\nu}$, is related to Riemann's curvature tensor $R_{\mu\alpha\nu\beta}$, and gives the curvature produced at each point in spacetime by the presence of matter and energy. Finally, $g_{\mu\nu}$, the metric tensor, gives all of the information about how the geometry of spacetime changes from point to point. It also represents the gravitational field itself, and that was a major discovery by Einstein. Three-dimensional space is another name for the gravitational field of the universe.

Ricci curvature tensor

Scalars, vectors and tensors

A mathematical quantity can be classified in terms of the number of components needed to describe it. For example, temperature and mass require only one quantity measurable in degrees celsius or in kilograms. These are examples of scalar quantities represented by T or M. Other quantities are more complex. For example the velocity of a car or the details of a magnetic force in space are examples of vector quantities that require three components in 3-D space: V = Vx,Vy,Vz, but can be reduced to a scalar quantity such as speed using the Pythagorean Theorem. Still other quantities such as pressure require not only a vector description but the component of this vector projected onto various surfaces in space. For example, a force normal to a surface projected in the x-y Cartesian plane would be represented as the pressure component P_{xy}. Pressure is an example of a quantity called a tensor. Other examples of tensors include gravitational fields, and strain within solid bodies.

The mass of a body warps the geometry of space to cause the force we call gravity.

What does one do with this equation? When you took algebra in school, you were asked to set up a problem and 'solve for x'. To solve Einstein's equation, you set up the problem by mathematically specifying where the matter and energy will be located throughout spacetime, $T_{\mu\nu}$. You then 'solve' the equation for $g_{\mu\nu}$, which will tell you, mathematically, what kind of geometry spacetime will have due to the amount and location of matter and energy within it. From $g_{\mu\nu}$ you can then calculate the exact worldline (geodesic curve) of any particle in spacetime, from light rays to rocket ships, by using the mathematical tools provided by Riemann.

Einstein's equation has been used to solve the simplest possible forms for $T_{\mu\nu}$ ever since he proposed it over one hundred years ago. Some of the 'trivial' solutions to this equation have led to the discovery of black holes (all the mass is at one point in spacetime) and the creation of Big Bang cosmology (matter is a dilute gas filling spacetime). Supercomputers are now programmed to do these formidable calculations to predict how black holes collide and merge as well as to predict other phenomena under real-world conditions.

Solving Einstein's equation

GRAVITY LENSES

Among the new predictions made by Einstein's General Theory of Relativity are that light rays can be bent as they travel through the warped spacetime near a massive body. If you use the Sun as the mass point viewed from Earth, the light from distant stars near the edge of the Sun appears to be deflected away from the Sun by a slight amount, which Einstein calculated to be about 1.5 arc seconds.

A foreground galaxy acts as a gravitational lens to bend the light rays from a distant galaxy behind it, distorting the image into multiple components.

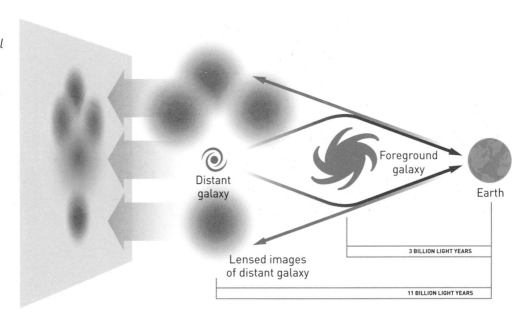

Distant galaxy

Foreground galaxy

Earth

Lensed images of distant galaxy

3 BILLION LIGHT YEARS

11 BILLION LIGHT YEARS

Albert Einstein

Perhaps one of the greatest physicists of all time, and certainly the most well-known in the 20th century, Albert Einstein had already made major contributions to physics before his 1905 publication of his Special Theory of Relativity. He explained the photoelectric effect in terms of Max Planck's theory of the quantization of light, and proved the atomic nature of matter by studying Brownian motion of dust in a drop of water. Either of these would have assured his stature in physics as a pre-eminent physicist. He was propelled into public view by the 1920 proof of one of his predictions of General Relativity. For his remaining years until his death in 1955, he made a long series of contributions to cosmology, quantum theory, and the search for a 'unified field theory' of physics, now called a Theory of Everything.

GRAVITATIONAL LENSING ▶ *objects with a large gravitational pull bend the light around them, and the larger the object, the greater the distortion.*

PROVING EINSTEIN WAS RIGHT

The total solar eclipse on 29 May 1919 was chosen as the test bed to decide whether Einstein or Newton was correct; the entire scientific world waited to hear the results. One of those chosen to carry out the test was the British astronomer, physicist and mathematician Arthur Eddington, who had championed Einstein's theories at a time when the German physicist was largely unknown in Britain. Eddington was a member of the expedition that sailed to the island of Principe, off the West Coast of Africa, where the total solar eclipse would be visible. A photograph taken during the eclipse showed that a star which would normally have been hidden behind the Sun was clearly visible, showing the bending effect of the Sun's gravity, as predicted by Einstein. At times other than during an eclipse, the glare of the Sun would have hidden this bending effect. This result was proclaimed by newspapers all over the world, including on the front page of the *New York Times* on 10 November 1919.

Eddington's expedition

Sir Arthur Eddington

This English physicist and astronomer left his marks on many areas of astrophysics. In 1920 he used a total solar eclipse of the sun to prove one of the cornerstone predictions of Albert Einstein's new Theory of General Relativity: the bending of starlight near the solar limb. In 1920, following many years of studying the mathematical properties of stellar interiors, he published his paper 'The Internal Constitution of the Stars' which presaged the discovery of the energy source of stars in nuclear fusion, through an application of Einstein's famous $E=mc^2$. He went on to investigate the stability of white dwarf stars and determined a mass limit for them of 1.44 solar masses based upon the principles of quantum mechanics and electron degeneracy pressure. Eddington even devised a gauge for measuring a cyclist's abilities called the Eddington Number such that a cyclist will cycle E miles in E days with E being the cyclist's ranking!

The calculation can also be performed using Newton's physics in which space is undistorted by the mass. But the amount of the predicted starlight shift would be exactly half as much because, under Newton, space remained unaffected by gravity and therefore not warped. So this was a very clear test of whether Newton was right, or whether a not-well-known Einstein was correct. The space around a star is not flat, as Euclidean maths suggests, but is warped as predicted by Einstein.

Starlight shift

An image from the Hubble Telescope shows an Einstein Ring.

This warping creates what is known as an Einstein Ring. The first was discovered by MIT astronomer Jaqueline Hewitt, who observed the radio source MG1131+0456 using the Very Large Array radio telescope. This observation saw a quasar lensed by a nearer galaxy into two separate but very similar images of the same object. The images stretched round the lens into an almost complete ring. The first complete Einstein ring, designated B1938+666, was discovered by collaboration between astronomers at the University of Manchester and NASA's Hubble Space Telescope in 1998. But even more complex gravitational lensing is possible if a lumpy mass distribution, like the galaxies in a cluster of galaxies, is used.

If so, the path taken by a single light ray will be more complicated than just a 1-dimensional angular deflection. The background object, a more distant galaxy behind the cluster, will have its multiple light rays bent into many different directions by the cluster, and form numerous distorted images of the same galaxy. The shape and distribution of these so-called 'lens arcs' can be used to determine the total gravitational mass of the cluster, and from a process called ray-tracing, be reassembled back into an undistorted image of the distant galaxy.

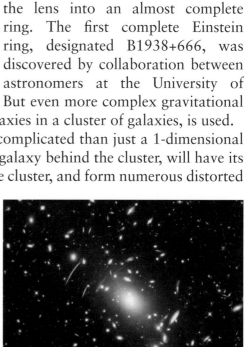

The blue arcs in this image of Abell 1063 are formed by the gravitational lensing of more distant galaxies behind this cluster.

By 1987, images of distant clusters of galaxies such as Abell 370 began to turn up odd images with an arc-like shape that did not look like any of the other galaxies in these clusters. In the mid-1980s, ground-based telescopic observations of the most prominent arc in this cluster allowed astronomers to deduce that the arc was not a structure of some kind within the cluster, but the gravitationally lensed image of an object twice as far away. Further studies and models of distant gravitational lens systems were not only able to reproduce the distortions in great detail using General

Relativity, but dramatically changed how astronomers studied the distant universe.

First, the gravitational lensing effect, together with a model of the masses in the cluster, could be used to reverse the distorting effects of the gravity to recover actual images of the lensed galaxies behind the clusters. Also, the lensing process amplified the light from these distant objects, allowing astronomers to glimpse galaxies forming in a much more distant, and therefore infant, universe. In some sense, the Hubble Space Telescope became the 'eyepiece' of a distant cosmic lens millions of light years across!

Secondly, the detailed mass model for the lensing cluster included all of the matter, invisible or otherwise, that was contributing to the gravity of the cluster. From this, not only the luminous mass for each cluster could be determined but also the 'dark matter' component could be weighed and its distribution throughout the cluster estimated. The issue of 'dark matter' will be discussed further in chapter five.

The lensing of background galaxies reveals the distribution of dark matter in a cluster.

Gravitational Lensing

The basic equation for the deflection of starlight by warped spacetime is given by

$$\theta = \frac{4GM}{Rc^2}$$

where G is Newton's constant of gravity (6.67×10^{-11} N m^2/kg^2), c is the speed of light (300,000,000 m/s), M is the mass of the object in kilograms, and R is the distance from the centre of the body to the point where the light ray passes the object (in metres). In the case of the Sun, if the star's image is located at twice the radius of the Sun from the Sun's centre, R = 1.4 million km (or 1.4×10^9 m) and M = 2×10^{30} kg, then

$$\theta = 4\ (6.67\times10^{-11})(2\times10^{30})/(1.4\times10^9)(3\times10^8)2 = 4.2\times10^{-6} \text{ radians}$$

Since 1 radian is equal to 206265 arsceconds of angular measure, the deflection of the star's image will be 0.9 seconds of arc.

GRAVITY WAVES

A further prediction from General Relativity was that, much like the electromagnetic field, gravitational fields can also support wave-like phenomena. In 1918, Einstein published a paper describing these waves and identified three different types. As a gravity wave passed by an observer, he or she would feel a specific pattern of metric changes in time as the distances in the local co-ordinate system changed in response to the distant, accelerating mass. In 1957, American physicist Richard Feynman showed that gravitational radiation could actually carry off energy from a system, which led to Joseph Weber creating the first gravity wave detector at the University of Maryland.

LIGO

Although he later claimed a detection in 1969, this was not independently confirmed by other gravity wave instruments. Nevertheless, the compelling argument that these waves existed led to decades of work on improving the sensitivity of detectors and leading to the Laser Interferometry Gravity wave Observatory (LIGO), whose construction started in 1994. The direct detection of gravity waves remained the Golden Fleece of physics for three decades, but it was known that something like gravitational radiation was at work to explain at least one well-studied astronomical system: the Hulse-Taylor Binary Pulsar (PSR B1913+16).

First discovered in 1974 by Russel Hulse and Joseph Taylor from the University of Massachusetts, this was the first pair of neutron stars found to be a binary system. One of these was its own pulsar, spinning dozens of times a second and emitting bursts of radio waves. From precise measurements of the pulsar's signals and timings, they deduced the details of the orbits and spins of the neutron stars to high accuracy. Over the course of the next decade, repeated remeasurements showed that the system was losing energy, which from gravitational radiation would amount to 7×10^{24} watts or some 2 per cent of the light emission from our Sun. The amount matched exactly what was predicted by General Relativity, even after corrections were made for energy lost by tidal distortion of the neutron stars themselves. At this rate, the neutron stars will spiral in and collide in about 300 million years. This collision process itself turns out to be an even more powerful source of gravity waves. Within less than one second, more energy can be released than is generated by all the stars in our Milky Way galaxy.

The Virgo gravity interferometer detector at the European Gravitational Observatory. It was used with the LIGO observatory to confirm gravity waves.

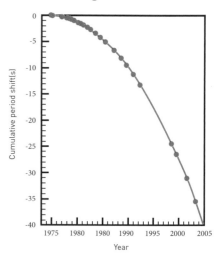

Energy lost by the orbiting Taylor-Hulse pulsars confirms the existence of gravitational radiation.

NEUTRON STAR ▶ *a very small, very dense star resulting from the gravitational collapse of a massive star after a supernova explosion.*

PULSAR ▶ *rapidly rotating neutron star or white dwarf that emits a very powerful beam of electromagnetic radiation. It was discovered by radio astronomer Jocelyn Bell Burnell.*

After years of producing null results, improved detectors became operational in 2015. LIGO made the first direct detection of gravitational waves on 14 September 2015. It was inferred that the signal, dubbed GW150914, originated from the merger of two black holes with 36 and 29 solar masses, creating a combined mass of about 62 solar masses. This suggested that the gravitational wave signal carried the energy of roughly three solar masses, or about 5×10^{47} joules. During the final fraction of a second of the merger, it released more than 50 times the power of all the stars in the entire observable universe combined. These black holes were not in our Milky Way back yard, but located 1.3 billion light years away.

Neutron Star collisions

Beyond studies of colliding black holes and neutron stars, gravity waves provide one of our deepest views into the physics of the early history of the universe. Gravity waves generated by the Big Bang event itself leave behind distinct fingerprints in the background radiation of space (cosmic background radiation, or CBR – see page 82), which can be deciphered to vastly improve what we know about events at the origin of spacetime itself.

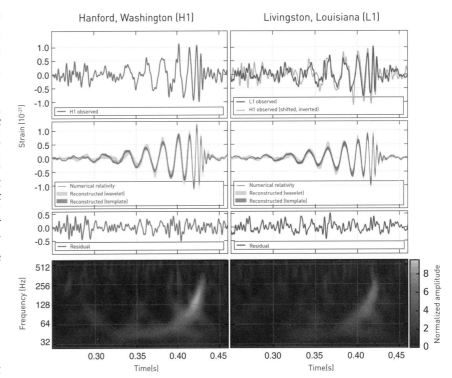

The gravity wave burst detected by the LIGO interferometer.

SPACE IS A 'MYTH'

General relativity provided the first glimpse of the intimate relationship between space, time and the gravitational field. Einstein's field equation showed much more than simply how gravitational forces could be thought of as the local curvature of spacetime. The mathematical quantity representing the metric of spacetime, $g_{\mu\nu}$ was absolutely identical to the field describing gravity. Gravity isn't just the curvature of spacetime, it *is* spacetime!

The issue of 'prior geometry' was the key to interpreting not only Einstein's choice of $g_{\mu\nu}$ to represent gravity, but in deciding whether $g_{\mu\nu}$ was actually a compound object in disguise; one part being the gravitational field, the other part representing a pre-existing and immutable arena of spacetime. To make such a decomposition work, the part of $g_{\mu\nu}$ that was prior geometry cannot be affected by matter or energy; that was the exclusive role to be played by the second component of $g_{\mu\nu}$ representing the gravitational field. Prior geometry would have to play the role of the absolute bedrock of spacetime from which both special relativity and Newtonian physics are built up, but it would have no physical effect upon matter or motion. No observation by the time Einstein proposed General Relativity, or since, has ever uncovered any physical evidence for some 'universal geometric object' or plenum which stands aloof from physics in the manner that prior geometry would have to.

Einstein's choice was that $g_{\mu\nu}$ represented *everything*, with no pre-existing framework for spacetime. This assumption, as provocative as it seems, is the simplest one that is consistent with all known phenomena and with the core ideas in relativity itself. Einstein once remarked that prior geometry *'is built on the a priori, Euclidean four-dimensional space, the belief in which amounts to something like a superstition.'* It is also implicit in the Leibniz relativistic view that space and time do not exist but are merely relationships between bodies, so that without bodies both time and space cease to exist.

Albert Einstein developed the theory of general relativity, which has proved to be essential to understanding the workings of the universe.

Chapter Four

RELATIVISTIC COSMOLOGY

Cosmic Expansion – de Sitter's Expanding, Empty Universe – Friedmann 'Big Bang' Solutions – Hubble's Law of Recession – Hot Big Bang Cosmology – Cosmic Microwave Background – Cosmic Horizons – Beyond the Horizon – CMB and the Horizon Problem

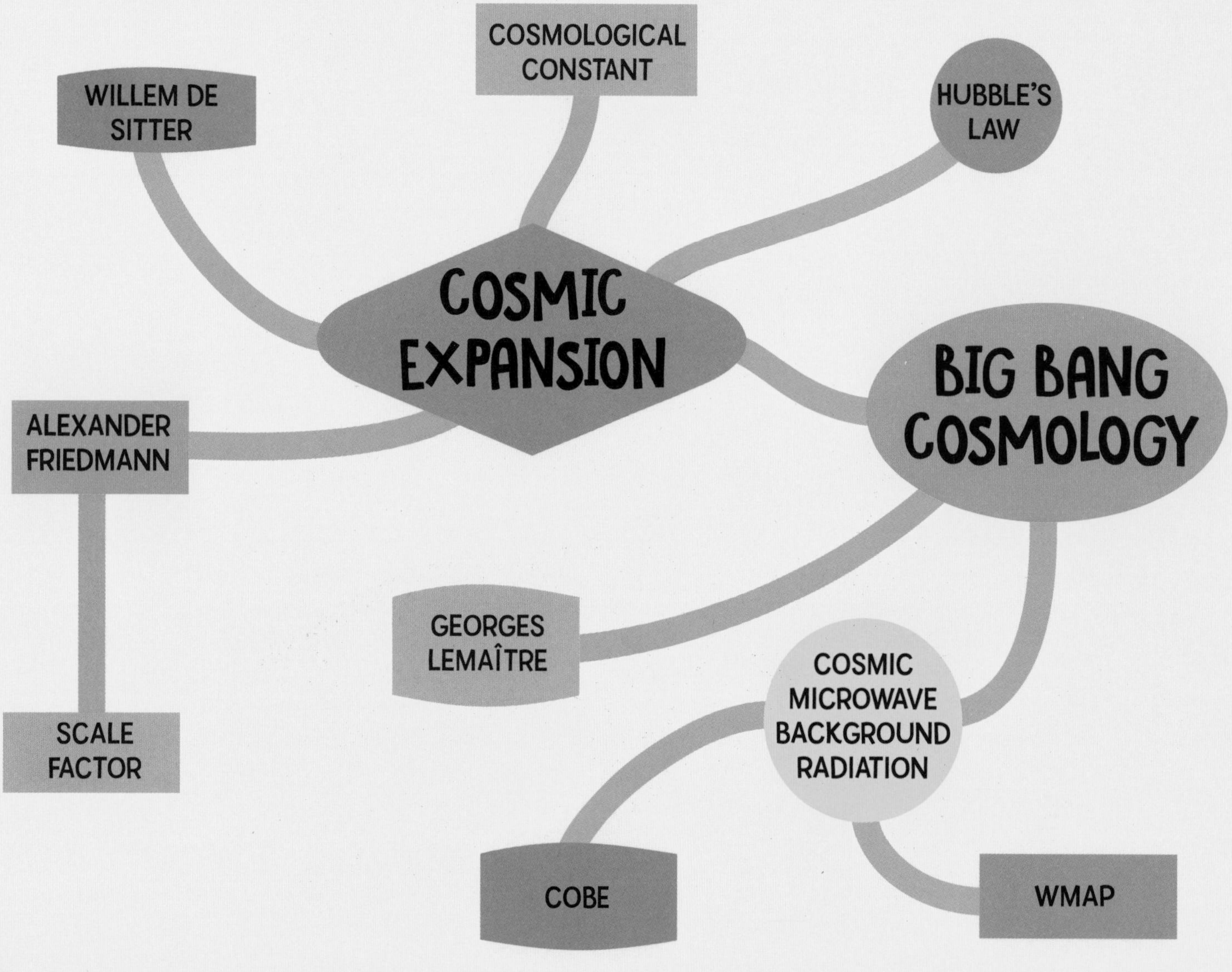

COSMOLOGICAL CONSTANT
WILLEM DE SITTER
HUBBLE'S LAW
COSMIC EXPANSION
BIG BANG COSMOLOGY
ALEXANDER FRIEDMANN
GEORGES LEMAÎTRE
COSMIC MICROWAVE BACKGROUND RADIATION
SCALE FACTOR
COBE
WMAP

COSMIC EXPANSION

Armed with a new relativistic theory for gravity and spacetime, Einstein immediately tried to solve the curvature equations to obtain a model for the cosmos. Frankly, no one had ever seen these equations before and so it would be Einstein himself who would have to set them up properly and then identify a solution to them. The left side of the equals sign was simply the mathematical expression for curvature in four dimensions, but the right side had to describe how matter and energy were distributed throughout spacetime.

If you selected a distribution where all the mass was at one point, you got a solution for 'black holes' (a region of spacetime producing such extreme gravitational effects that not even light can escape – see page 149). But if you decided that matter was evenly distributed everywhere, called '*isotropic* and *homogeneous*', you got a second family of solutions, and these were quite fascinating. *Isotrophy* means that what you observe around you when viewing from one location looks the same, no matter in which direction you look (*rotationally invariant*). *Homogeneity* means that what you observe around you looks the same, even when you change your location (*translationally invariant*).

Distribution of matter

ISOTROPHY ▶ *from the standpoint of any observer, the principle that matter seen across the two-dimensional sky looks uniform in angular degree.*

HOMOGENEITY ▶ *the principle that matter has a uniform distribution along the third dimension out into the depths of space.*

Einstein had considered what was known about the universe at that time, around 1917, and decided to favour Isaac Newton's proposition that matter was smoothly distributed across all of space, and could be represented by a simple value like the ordinary density of gas. Never mind that this cosmological gas represented all of the stars within galaxies dissolved into a gas with an average density of about 11 atoms per cubic metre. When he used this value on the right side of the equation, and assumed *homogeneity* (density doesn't depend on distance variables like x, y, z or r) and *isotrophy* (density doesn't depend on direction variables like q or f), this caused a substantial reduction in the number of equations needed on the left side to define curvature. But the result was not what he expected. The solution required that the universe be in a state of collapse, which completely disagreed with what he thought astronomers were saying about the stars and nebula in the sky.

Cosmological gas

To prevent the whole universe collapsing and remain more or less static in time, Einstein felt compelled to add a *cosmological constant* – a 'fudge factor' – represented by the Greek letter Λ to his equation for gravity as shown below.

Fudge Factor

$$R_{\mu\nu} - {}^{1}/_{2}R\, g_{\mu\nu} + \Lambda\, g_{\mu\nu} = \frac{8\pi G}{c^{4}}\, T_{\mu\nu}$$

Cosmological constant

This factor could be exactly selected to make the universe stable. Einstein's cosmological constant term was extremely strange. What it said was that hidden within space, within every cubic metre out to the farthest scrap of matter, there was a peculiar 'something' that created an anti-gravity force in nature. By some miracle, the amount of this force at the local scale of the solar system was precisely enough to allow the entire universe to remain fixed in time. Einstein had no idea whether this was a plausible phenomenon, especially since astronomers had as yet no way to measure the speeds of distant cosmic matter, nor could he firmly declare that such matter even existed. It is not even clear that Einstein was aware of the research that was available into the nature of extragalactic nebulae or the motions of stars via the Doppler Effect. Both pointed towards a dynamic and changing universe. But other physicists became very excited by Einstein's General Relativity, especially its cosmological predictions, and went considerably further than Einstein in exploring his own equations for curvature.

COSMOLOGICAL CONSTANT ▶ *the energy density of space which allows the universe to remain fixed in time.*

DE SITTER'S EXPANDING, EMPTY UNIVERSE

First, in 1917, the Dutch astronomer, physicist and mathematician Willem de Sitter described an expanding cosmos devoid of matter, which was an extreme solution to Einstein's equations with vanishing density and a non-zero cosmological constant. Einstein did not like that solution because if you introduced just two test particles, they would move apart from each other over time at an exponential pace set by the value of the cosmological constant. What was worse, these universes devoid of any matter were not at all static.

In 1922, the Russian mathematician Alexander Friedmann found solutions to the equations when the geometric radius of curvature, or scale, of the universe was allowed to change in time with the change in the density of matter. These 'Friedmann' solutions each gave a specific formula for how the 'scale factor' of the universe would change in time given specific values for the matter density and the cosmological constant.

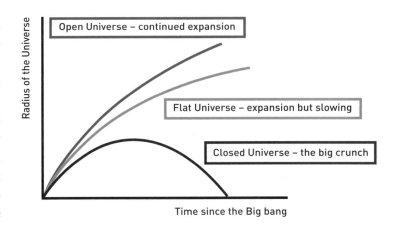

If you set the matter density to zero and left the cosmological constant intact, you got de Sitter's empty universe expanding in time, and as Einstein had deduced, the separations of a few particles would increase exponentially. If you set the cosmological constant to zero, however, you got solutions in which the universe continued to expand, but now you could actually calculate from the matter density just how the overall geometry of the universe would change, and how the speeds of matter points would change with time based upon the matter density.

One solution described the geometry of the universe as closed and finite with a positive and constant curvature, in the same way that the surface of a sphere is a closed surface of finite area. These universes expand to a maximum size and then re-collapse. Then there were two solutions that predicted open, infinite universes: One had a flat spacetime geometry with zero curvature like the spacetime of Special Relativity. The other had a hyperbolically shaped geometry with a negative constant curvature.

In all cases, the density of the universe and its current critical density determined which universe we lived in. If the density was too high compared to the expansion speed, the universe would be a closed and finite one destined to re-collapse in the future. If the density was at a critical value, the universe would have a flat geometry with exactly zero curvature. If it had too little density, it would have a negative-curvature geometry.

THE FRIEDMANN 'BIG BANG' SOLUTIONS

The Friedmann solutions were mathematical formulae that described how the scale factor of the universe changed in time. In other words, this scale factor was a measure of how the distances between any two objects increased.

An important, though counter-intuitive feature of General Relativity and Big Bang cosmology involves the expansion of space, or more correctly, the increase in the separations between worldliness (the path that an object takes in both space and time). We encountered the relativistic concept of the worldline in Chapter 3 when we discussed the idea of spacetime. The Friedmann solutions refer to a scale factor that changes with time. What this refers to is a new concept of

The Cosmic Scale Factor

To understand the important role of the cosmological scale factor, a(t), let's consider an analogy: the locations of Paris and New York. The latitude and longitudes of these cities are fixed and the distance between them is related to the radius of Earth. Now suppose we defined this radius as R = a(t) × 6,378 km so that today the scale factor a(now) = 1.0. Then you could calculate the distance between these cities and get the usual amount. But suppose for some reason the mass of Earth was declining over time so that a(t) was not always 1.0. Then the distance between New York and Paris would change over time even though their latitudes and longitudes remained fixed. What the Friedmann solutions provided was a prediction for what this scale factor a(t) looked like. The so-called Big Bang solutions required that a(t) was increasing in time at a rate determined by the current density of the universe compared to a critical density; a ratio defined by the symbol Ω. This meant that:

- If the density were higher than critical ($\Omega > 1.0$), the universe would re-collapse and have a closed geometry with a(t) reaching a maximum value and then decreasing.
- If the density were lower than critical ($\Omega < 1.0$), the universe would continue to expand indefinitely and a(t) would increase without limit.

Even though the 'latitude and longitude' of the galaxies remained the same, their physical separations changed over time. Also, because a(t) was a property of space and not matter or information, its rate of change was not limited by the speed of light. In General Relativity, the separations between cosmological objects can increase faster-than-light even though the objects embedded in space such as galaxies and light rays, cannot do so.

An expanding balloon demonstrates cosmic expansion.

motion that is entirely non-Newtonian. In Newton's physics, objects are located at specific co-ordinates in a fixed three-dimensional space. Movement means that the objects change their separations between each other in a unit time by moving from one set of co-ordinates to another. The application of the Pythagorean Theorem based on co-ordinate differences then defines the distance travelled.

Critical Density

$$\rho = \frac{3H^2}{8\pi G}$$

A major success of Big Bang cosmology is that it relates the average density of the universe to its rate of expansion measured by Hubble's Law (H). The current best measured value for H = 71 km/s/mpc or 1/9.77 billion years. With G = 6.6 × 10^{-11}m^{3}kg^{-1}s^{-2} and converting H into seconds to get H = 3.3 × 10^{-18} s^{-1} we get r = 1.9 × 10^{-26} kg/m^3. This is equal to 11 hydrogen atoms in every cubic metre of space in the visible universe. To decide between the three possibilities for the universe described by the Friedmann solutions, you just compare the observed density of matter in the universe against the critical density – a ratio defined by the symbol Ω.

In General Relativity, cosmological motion is more complicated than this. Galaxies are located at fixed co-ordinates in space, but the separation between them changes because space itself dilates. The reason that distant galaxies appear to be receding from us at tremendous speeds is not because they are moving close to the speed of light, but that the space between them and our location has dilated considerably as the universe has expanded during the intervening time period. This is, intuitively, a very difficult idea to understand, much like how electrons can behave as wave-like or particle-like objects, or that high speed travel causes clocks to get pulled out of sync due to the 'time dilation' effect. There is no intuitive understanding of these observed phenomena, yet we are forced to accept them due to the preponderance of data supporting them.

Cosmologically, every galaxy is located at a fixed set of co-ordinates, but the distances between them are determined by the value for a(t). According to General Relativity, a(t) is not a physical thing, so it is free to change in time such that the distances between galaxies can increase faster than light, and they did so during the early moments of the Big Bang. This is also consistent with Einstein's relativity, which states that what we call space is an illusion. Galaxies do not move *through* space, but in some sense are dragged along by the dilation of space. This dilation is only discernable in large enough reference frames where Special Relativity no longer holds and you need General Relativity to define the properties of spacetime

The Belgian astronomer and physicist Georges Lemaître, who was

Alexander Friedmann, the man responsible for developing the big bang solutions.

also a Roman Catholic priest, took a look at Einstein's equations independently of Friedmann in 1927 and made the discovery that it was possible from the motions of distant bodies to determine in which universe you lived. Specifically, you could tell whether the universe was static, expanding or contracting as the various Friedmann solutions indicated, though Lemaître did not know about Friedman's work. For an expanding universe, there was a simple relationship between the measured line-of-sight recessional speeds measured by Doppler techniques and the distance to the object according to V = Hd. In this equation, V is the recession speed of a galaxy as viewed from Earth, d is its distance from Earth and H is a constant that defines the rate at which the scale factor in Friedmann cosmology is changing. Lemaître estimated the value to this constant, H, and published his results in 1927. He is generally credited as being the Father of Big Bang Cosmology for his V=Hd discovery.

Edwin Hubble developed the Hubble Constant.

HUBBLE'S LAW OF RECESSION

By 1929, the American astronomer Edwin Hubble, working at the Mount Wilson Observatory, had put together the Doppler and distance data for 46 galaxies and plotted them on a distance v speed diagram, from which he deduced a linear trend. For every million parsecs in distance, the recession speed would increase by about 500 km/s. This rate is now called the Hubble Constant and represented as H_0 in the formula $V = H_0 d$.

Astronomical Distance Units

Astronomers use several different measures of interstellar and intergalactic distances. The oldest of these is the light year, first mentioned in 1838 by the German astronomer, mathematician and physicist Friedrich Bessel and later by the German science writer Otto Ule in an 1851 popular astronomy article. It is the distance that light travels in exactly one Earth year – a distance of 9.46 trillion km. Next came the astronomical unit (AU), the mean distance between the centre of the Earth and the centre of the Sun, first mentioned in 1903. It is 149 million km. The parsec came into use around 1913 and is an abbreviation of parallax arc second (par-sec). It represents the distance at which the radius of Earth's orbit (AU) subtends an angle of exactly one second of angular arc (1/3600 degree). It is equal to 20,6265 AU or 3.26 light years, or equivalently 30.85 trillion km. Astronomers tend to use astronomical units when comparing distances within the solar system; light years when describing the sizes of objects such as nebulae, star clusters and galaxies; and parsecs – or specifically mega-parsecs – when describing cosmological distances.

Another feature of the Friedmann solutions is that the Hubble Constant could be used to estimate the age of the universe. For infinite universes with a flat geometry, $T = 2/3Ho$. For Hubble's original estimate of 500 km/s/megaparsec, and with 1 megaparsec = 3×10^{19} km, one gets an age of 4×10^{16} s or 1.3 billion years. But over time, and as better data accumulated, the value of the Hubble constant has been greatly reduced so that the current high-precision age estimate is closer to 14 billion years.

Hubble constant

But what of Einstein's cosmological constant, Λ, which was not a part of the Hubble Law prediction? In 1930, Arthur Eddington published a paper in which he showed that Einstein's static universe with its cosmological constant was actually unstable, and so it would not have helped Einstein at all. That being the case, it could be eliminated from cosmological theories and the only solutions to Einstein's equations were the three solutions found by Friedman with only matter in the universe.

In 1932, Einstein and de Sitter published the Einstein–de Sitter universe, which became the standard model up to the middle of the 1990s. It is a spatially flat, ever-expanding universe with $\Lambda = 0$ and an expansion velocity approaching zero in the infinite future. They had effectively kicked the cosmological constant completely out of their cosmological equations. Once Einstein knew the universe was expanding from Hubble's measurements, he discarded the cosmological constant as an unnecessary fudge factor. He later called it the 'biggest blunder of my life', according to fellow physicist George Gamow. The cosmological constant has since been resurrected as a possible explanation for a mysterious force discovered in the 1990s called 'dark energy' (see pages 99–100).

Einstein's conundrum

The *New York Times* on 11 February 1931 included an interview with Einstein: *'The redshift of distant nebulae has smashed my old construction like a hammer blow...The red shift is still a mystery. The only possibility is to start with a static universe lasting a while and then becoming unstable and expansion starting, but no man would believe this... A theory of an expanding universe at the rate figured from apparent velocities of recession of nebulae would give too short a life to the great universe. It would only be ten thousand million years old, which is altogether too short a time. By that theory it would have started from a small condensation of matter at that time.'*

NAMES TO KNOW:
BIG BANG COSMOLOGY

Willem de Sitter (1872–1934)

Albert Einstein (1879–1955)

Alexander Friedmann (1888–1925)

Edwin Hubble (1889–1953)

Georges Lemaître (1894–1966)

George Gamow (1904–1968)

THE HOT BIG BANG COSMOLOGY

In 1931, Georges Lemaître proposed in his 'hypothèse de l'atome primitif' ('hypothesis of the primeval atom') that the universe began with the 'explosion' of the 'primeval atom' – what was later called the Big Bang in the 1950s by British astronomer Fred Hoyle, who had a competing idea called 'Steady State Cosmology'. Lemaître had to wait until shortly before his death to learn of the discovery of cosmic microwave background (CMB) radiation: the actual remnant radiation of Lemaître's dense and hot phase in the early universe.

George Gamow led the development of the 'hot Big Bang' theory in the 1940s. He was the earliest to employ Alexander Friedmann's and Georges Lemaître's non-static solutions of Einstein's gravitational equations to explore what happens to matter during the early history. By 1946, Gamow assumed that the early universe was dominated by radiation rather than by matter. Working out how nucleosynthesis would proceed led to the prediction of an intense radiation field of cosmic light whose current temperature would be about 20 K. Gamow also found that it was impossible to create the other elements in the periodic table heavier than helium.

George Gamow developed the 'hot Big Bang' theory.

NUCLEOSYNTHESIS ▶ *the formation of new atomic nuclei from protons and neurons.*

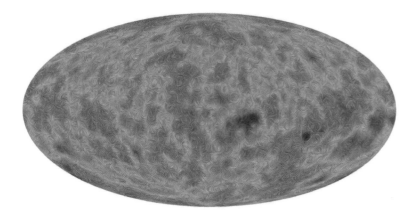

A visualization of the Cosmic Microwave Background as detected by the European Space Agency's Planck satellite.

COSMIC MICROWAVE BACKGROUND

It was recognized early on that the initial 'fireball' light from the Big Bang would start out as a very high-temperature light at gamma ray energies and higher, but as the universe expanded and cooled, this light would maintain a *black body* spectral shape (that is, defined only by its temperature), although its peak energy would steadily shift to longer wavelengths. By the mid-20th century, scientists knew that the light would be detectable only dimly at microwave radio wavelengths at a temperature only a few degrees above absolute zero. It remained to successfully detect this important prediction by Big Bang cosmology.

Discovery of the Cosmic Fireball Light

In 1964, radio engineers Arno Penzias and Robert Wilson at Bell laboratories were trying to reduce the radio interference ('noise') in their microwave antennae but could not entirely eliminate it. Its temperature was about 4.2 K. They contacted physicists Robert Dicke and David Wilkinson at Princeton University, who were trying to detect the cosmic microwave background radiation signal, and the news was met with considerable excitement. Dicke persuaded Penzias and Wilson to publish their results in *The Astrophysical Journal*, along with the suggestion by Dicke and his team that it was evidence of CMB. Because astronomer Fred Hoyle's competing theory to 'Big Bang' cosmology called Steady State Cosmology also offered an explanation for this radiation, the Cosmic Microwave Background (see opposite) was not firmly established as proof of Big Bang cosmology until the 1970s when further measurements showed it to be a black body spectrum – a feature that only Big Bang cosmology also predicted.

'Well, boys, we've been scooped!' – how physicist Robert Dicke broke the news to his team at Princeton after hearing that the CMB signal they had been searching for had been detected by researchers at Bell Laboratories.

Since the 1970s, continued work using satellites such as NASA's Cosmic Background Explorer (COBE), the Wilkinson Microwave Anisotropy Probe (WMAP) and ESA's Planck spacecraft have measured the CMB to high accuracy. The first historic measurement of the precise temperature of the CMB was made by the COBE Far Infrared Absolute Spectrophotometer (FIRAS) team led by NASA astronomer John Mather. Their initial 10-month 'quick look' results were announced at the January 1990 meeting of the American Astronomical Society. Mather, along with physicist George Smoot, went on to receive the 2006 Nobel Prize in Physics for the COBE results.

They not only verified its constant temperature near 2.725 K and its near-perfect black body spectrum, but that it is largely isotropic across the sky to better than one part in 1000. However, at still higher precision, variations in its temperature at 1 part in 100,000 can be seen that represent the cosmic background radiation (CBR) interacting with the irregular, clumpy matter that existed within 380,000 years after the Big Bang. These irregularities led to further refinements on the Big Bang model, including evidence for dark matter, dark energy and a period of rapid expansion called 'inflation' (discussed later). The detailed investigation of these phenomena make up the bulk of cosmology in the 21st century.

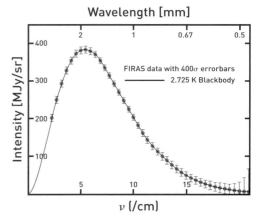

The black body spectrum of the CMB discovered by COBE.

John Mather

John Mather is an American astrophysicist born in 1946 in Roanoke, Virginia. He received his PhD in physics at the University of California at Berkeley in 1974, and went on to work on the cosmic background explorer concept, first at the Goddard Institute of Space Studies from 1974 to 1976, then at the NASA Goddard Spaceflight Center in Maryland from 1976 until the COBE spacecraft launched in 1988. He has been the NASA Project Scientist for the James Webb Space Telescope since 1998, due to launch in 2021. In 2007 he was listed by *Time* magazine as one of the 100 most influential people in the world.

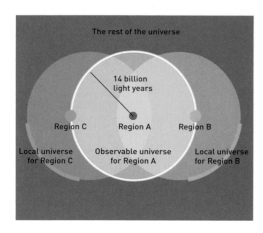

An example of cosmic horizons in Big Bang cosmology.

COSMIC HORIZONS

General Relativity says that there is no universal definition for speed that resembles the Newtonian one based on our common sense ideas about motion. Cosmological motion via space dilation and scale-factor change leads directly to Big Bang cosmology, and explains both the high-speed recession of distant galaxies, along with the current temperature of the cosmic background radiation. That being the case, it also leads to the existence of what are called horizons in the cosmos.

COSMOLOGICAL HORIZON ▶ *the 14-billion light year limit from which information can be observed.*

The consequence of horizons is that from every location in space there are places we can see today, and more distant places we will never see until the universe ages. Horizons grow in time, and eventually in a universe that expands into the infinite future, all of these horizons will overlap completely and each observer will have then seen every object in the universe which emerged from the Big Bang. Before then, you can have three observers, call them A, B and C, arranged in the cosmos such that A can now see B, and B can now see C, but A cannot see C. It is easy to draw a diagram on a piece of paper that shows how the distances have to be arranged between A, B and C relative to the current horizon distance of 14 billion light years to make this logic work.

BEYOND THE HORIZON

What does the universe look like beyond our current horizon distance? Because Big Bang cosmology is based upon the idea that matter is uniformly spread throughout space (recall the ideas of homogeneity and isotrophy), the little corner of the universe we can see out to our horizon should at least statistically look very much like what any other observer might see who lives beyond our horizon limit. For example, as they look in our direction some 12 billion light years from their location, they will see images of objects that look very much like the ones we see as we look to the edges of our own horizon and visible universe, except that one of the infant galaxies they will be seeing in images of 'today' will be images of our Milky Way as it looked 12 billion years ago.

Without any evidence other than the principle of uniformity and that our vantage point is not a special one in this universe, astronomers generally believe that every observer in the larger universe of which we are a part is seeing the same kinds of objects and also observing the same kinds of physical laws as we are. Everyone may see quasars (see page 145) at distances of several billion light years, but not necessarily the same quasars.

CMB AND THE 'HORIZON PROBLEM'

When we look the other way and ask what happens to these horizons at earlier times in the history of the universe, we run up against a severe problem. If two objects today are just able to signal to each other using light because they are inside each other's current horizons, looking back only one billion *Horizons* years, neither of these two objects was in contact with each other when the universe was only 13 billion years old. Following the Big Bang back in time, the matter out of which the Milky Way and the Andromeda Galaxy were formed was just dense clouds of matter separated by about ten billion kilometres when the universe was about one second old. But light could only have travelled one light second or 300,000 km in that time, and so at this time these two great masses in the cosmos would not have felt each other's light or even gravitational effects.

Although this seems very extreme, this horizon effect is something that could be tested using the CMB. The radiation that we map across the sky seems to be smooth and uniform, which *Uniform radiation* means the matter it was last in contact with had the same temperature to one part in 100,000. But if we look at when this radiation was last in contact with matter, it was when the universe was about 380,000 years old. Normally, two objects reach the same temperature by exchanging heat radiation (infrared light) which travels at the speed of light. But 380,000 years after the Big Bang, light could only have travelled 380,000 light years, which defined its horizon at that time.

At the present time, this distance across the sky corresponds to about 0.8 degrees. This means that if you pick two spots in the sky that are more than twice the diameter of the full Moon (0.5 degrees) from each other, the temperature of the CMB you measure should be more and more different the farther apart the points are. This is not at all what we actually see, and so by some means, the temperature of the matter some 380,000 years after the Big Bang was almost exactly the same everywhere. This uniformity in the face of cosmological horizon limitations was called the 'Horizon Problem', and it is a feature of all Big Bang cosmologies described by General Relativity.

HORIZON PROBLEM ▶ *the theory that all areas of the universe have maintained a uniform temperature despite not being in contact for nearly 14 billion years. Cosmic inflation is the currently accepted explanation for this.*

THE COSMOLOGICAL SINGULARITY

All Big Bang theories that have a finite origin in time predict a singular state at time = 0 when the scale factor of the universe reaches zero. This means the separations between all physical objects vanish and time literally ceases to exist. But current space contains matter so the density of the universe (mass divided by the volume of space) becomes infinite and so does the strength of the gravitational field. This state is called the Cosmological Singularity, and is a feature of all cosmologies based upon General Relativity. Stephen Hawking and Roger Penrose, moreover, proved in the 1960s that the worldlines of all particles in the cosmos today could not avoid being terminated at this singularity because of the collapse of space.

A variety of mathematical studies of the Cosmological Singularity were pioneered by Russian physicists leading to the idea that matter could be produced by the intense fluctuations in the gravitational field near that time. The details for how specific kinds of particles such as protons and electrons were created was not proposed because these models were rooted in general relativity rather than quantum mechanics. All that was really known was that there were several classes of Big Bang cosmology based on their spatial symmetries of which many could be highly anisotropic near the Cosmological Singularity. They were called Mixmaster Cosmologies to indicate the violent anisotropic changes that could take place in time. From simple $E=mc^2$ energy considerations, gravitational fluctuations seemed to offer a free energy source for creating matter.

These studies have now been superseded by applying quantum gravity techniques to describing what happens to spacetime during the Planck Era (the earliest stage in the birth of the universe, immediately after the Big Bang). Some studies suggest that the Singularity is avoided by new quantum 'antigravity' forces. Others describe the event as being smeared out by Planck-scale limits to the minimum scale of spacetime and limits to the maximum possible density (10^{94} gm/cc) and temperature (10^{32} K).

Stephen Hawking greatly expanded our understanding of the Cosmological Singularity through his research on worldlines.

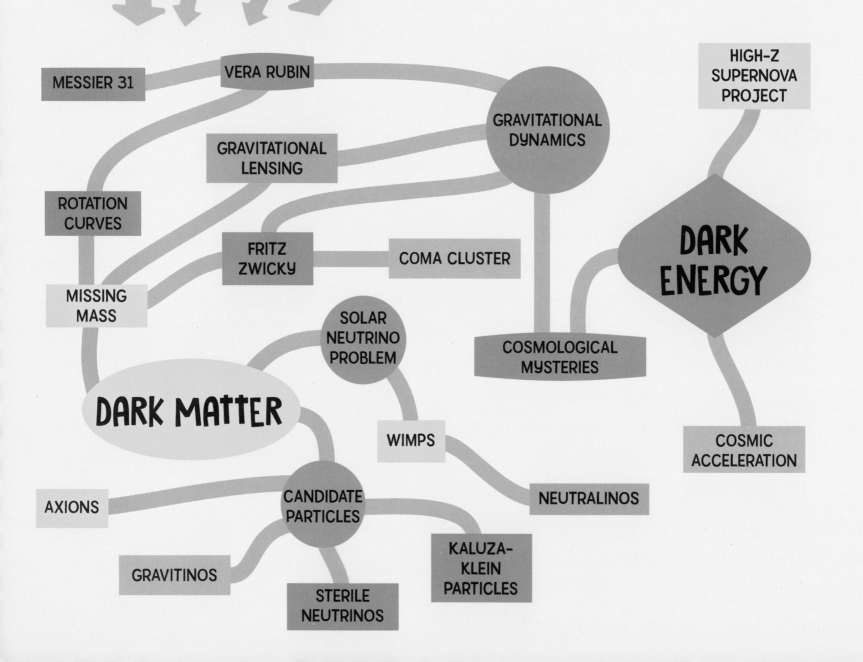

MESSIER 31

VERA RUBIN

HIGH-Z SUPERNOVA PROJECT

GRAVITATIONAL DYNAMICS

GRAVITATIONAL LENSING

ROTATION CURVES

DARK ENERGY

FRITZ ZWICKY

COMA CLUSTER

MISSING MASS

SOLAR NEUTRINO PROBLEM

COSMOLOGICAL MYSTERIES

DARK MATTER

WIMPS

COSMIC ACCELERATION

AXIONS

CANDIDATE PARTICLES

NEUTRALINOS

GRAVITINOS

KALUZA-KLEIN PARTICLES

STERILE NEUTRINOS

DISCOVERY OF MISSING MASS

In addition to intensive studies of the cartography of how galaxies are distributed throughout extragalactic space, a variety of studies of the gravitational dynamics of individual galaxies and clusters of galaxies have also taken place. These date back to the 1930s with the investigations by Caltech astronomer Fritz Zwicky, and further studies by Vera Rubin at the Carnegie Institute in Washington DC in the 1970s.

Galaxies in the core of the Coma Cluster move too rapidly for the cluster to be stable unless dark matter is present.

Zwicky studied the galaxies in the Coma Cluster in 1933, and systematically measured Doppler velocities for a small number of them. What he found was that the variation of their speeds within the cluster was over 2000 km/s. This would not be surprising in itself for a cluster containing over 800 galaxies, but a detailed calculation of their predicted speeds yielded a much smaller dispersion of less than 100 km/s. When he ran the calculation in the other direction to determine the average masses for 800 galaxies within a volume one million light years across, the amount of missing mass, or 'dark mass' as he called it, amounted to 45 billion Suns.

Fritz Zwicky, standing in front of the Schmidt Telescope on Palomar Mountain, California in 1937.

Fritz Zwicky
Zwicky was born in 1898 in Bulgaria, and received his advanced education in mathematics and theoretical physics at the Swiss Federal Polytechnic before immigrating to America in 1925, where he worked with physicist Robert Millikan at the California Institute of Technology. By 1933, his studies of the Coma cluster were the first to use the Virial Theorem to 'weigh' the clusters based on the measurable velocities of member galaxies. He identified the 'missing mass' as *dunkle Materie*, or 'dark matter' – the first use of this term, cosmologically.

This amount of mass would be over 200 times the estimate based on the normal light output from so many stars for which a typical mass-to-light ratio (M/L) is about M/L=3 rather than the M/L= 800 found by Zwicky. Our Sun has a mass of 2×10^{33} g and a luminosity of 3.8×10^{33} ergs/s, so its M/L is about 0.5 by comparison. To make the large velocity dispersion measured for Coma consistent with the understanding of galactic physics, there had to be an enormous amount of faint or dark material present in nearly every galaxy which was not being registered by simply counting the mass in the luminous stars. So much additional mass also wreaked havoc with the motions of stars in galaxies, which would now have to be dramatically different and moving at far higher speeds to be consistent with so much additional gravitational force.

Mass-to-light ratio

Another possibility was that the Coma Cluster was not a stable group of galaxies but was a transitory, ephemeral grouping destined to last a few billion years or less. However, Zwicky's studies of other clusters found the same missing mass problem, and it seemed statistically unlikely that we would be living at a time when so many clusters had 'magically' come together.

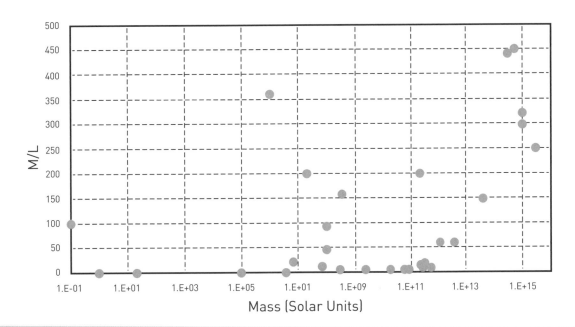

MASS–TO–LIGHT RATIO ▶ *a figure derived by dividing the mass of a star, galaxy or cluster by its luminosity. By identifying a star type and measuring its luminosity, you can work out its mass.*

CHAPTER 5

Vera Rubin conducted pioneering research into the rotation of galaxies.

Vera Rubin

She was born Vera Florence Cooper in 1928 in Philadelphia, but the family moved to Washington DC in 1938. She avoided her High School science teacher's advice to become an artist and instead pursued her passion for astronomy at Vassar College. At Georgetown University, her doctoral advisor was George Gamow. As a staff member at the Department of Terrestrial Magnetism, she did much of her work at home while raising her children. In 1963 she began a long collaboration with Geoffrey and Margaret Burbidge to study the rotation of nearby galaxies. She went on to discover the anisotropic expansion of the universe at scales of 100 million light years – known as the Rubin-Ford Effect. She is also, specifically, known for her discovery of massive, dark haloes of galaxies via studies of their rotation.

ROTATING GALAXIES

On the local scale of individual galaxies, Vera Rubin had studied the rotation speeds of stars and nebulae in several nearby galaxies such as Messier 31 in Andromeda and had discovered a curious effect. If most of the mass of a galaxy were concentrated in the visible disk of the galaxy, one could predict from Newtonian gravity that the rotation speeds should at first rise to a maximum, and then fall steadily according to Kepler's Law as you moved farther from the central concentration of light in the disk. Instead, what was seen by Rubin from detailed spectroscopic studies of galaxy 'rotation curves' was that the speeds remained very high even to the edge of the visible disk.

ROTATION CURVE ▶ *the orbital speed of stars in a galaxy plotted against their distance from the galaxy's centre.*

Radio astronomers, meanwhile, had been studying the distribution of interstellar hydrogen gas in a variety of spiral galaxies, including the Milky Way, to measure the speed of the rotation of the hydrogen gas. Although hydrogen atoms emit a specific 21 cm spectral line detectable by radio telescopes, the doppler shift of this line could be used to track the speed of hydrogen gas clouds far beyond the visible light disk of a galaxy. The radio astronomers discovered that hydrogen clouds detected beyond the visible stellar disk continued to have very high speeds.

The only simple explanation that gave such curves was that the entire galaxy was embedded in a halo of unseen matter, and that when this was figured into the mass-to-light ratio for a galaxy, the numbers would exceed 100 or more – just as had been surmised in the 1930s by Fritz Zwicky. The problem of *dark matter*, as it was now called, was now more than an issue for the stability of clusters of galaxies, but also applied to individual galaxies. Without this dark matter, the speeds of galaxies in clusters would be insufficient to stabilize them as cosmic structures. For galaxies, the individual stars would be moving too fast for the stellar mass in the galaxy to hold the system together for more than a few 100 million years.

The various components that contribute to the rotation curve of the Milky Way, showing the effect of dark matter.

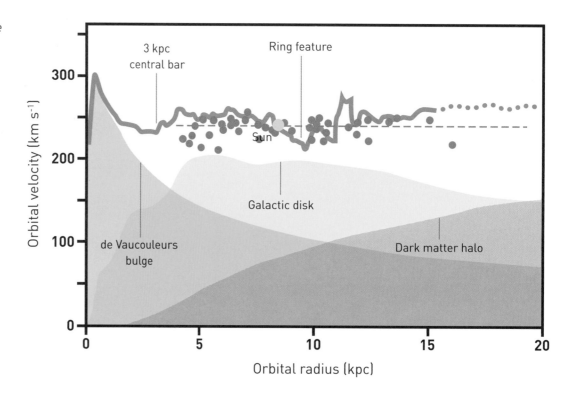

THE DARK MATTER MYSTERY

Since the time that Fritz Zwicky and Vera Rubin discovered 'missing mass' in clusters of galaxies and in the halos of individual galaxies, the notion that the universe contains gravitating matter not detectable through its luminous properties has grown to become one of the major curiosities and compelling mysteries of modern cosmology. Astronomers continue to identify more clusters and individual galaxies with this plague of 'dark matter', and eliminate many possibilities such as dim stars, black holes and a hot intergalactic medium.

DARK MATTER ▶ *a significant contribution to the gravitating mass of the universe not in the form of known types of matter such as electrons, protons, photons or neutrinos.*

A rendering of where dark matter may exist in the Milky Way.

The kinds of particles responsible for this dark matter on the cosmological scale also had to have certain properties based on supercomputer modelling of the evolution of structure in the universe. If the particles were too light (called *hot dark matter,* or HDM), they would wash out much of this structure, leaving behind very few clusters of galaxies and superclusters of galaxies by the present age of the universe. If the particles were too massive, they would be too effective in creating small-scale structures, which again would not resemble the patina of clusters strewn across intergalactic space. The term *weakly interacting massive particles* (WIMPS) caught on in the mid-1980s, when astronomers David Spergel and William Press suggested that they might also solve the Solar Neutrino Problem (see box, page 97).

WEAKLY INTERACTING MASSIVE PARTICLES (WIMPS) ▶ *the particles it is believed make up dark matter.*

This image from the Hubble Space Telescope shows the inferred location of dark matter (shown in blue) in the distant galaxy cluster Abell 1689.

Dark Matter in the Milky Way

Kepler's Third Law can be rewritten using Newton's Law of Gravity to serve as a means for 'weighing' any object that has an orbiting satellite:

$$V^2 = \frac{GM}{R}$$

V is the orbit speed of the satellite, M is the mass of the object being orbited and R is the radius of the orbit. For the Sun orbiting the centre of the Milky Way, R = 27,000 light years (2.6×10^{20} m), G is Newton's gravitational constant (G = 6.67×10^{-11} N m^2/kg^2) and V is the speed of the Sun in its orbit around the galactic centre, which is about 220 km/s (2.2×10^5 m/s). Solving for M and inserting the numerical values, you get M = 1.9×10^{41} kg. The mass of our Sun is 2.0×10^{30} kg, so the mass interior to the Sun's orbit must be about 100 billion solar masses. But careful studies of the motions of stars and nearby galaxies beyond our Sun's orbit shows that our galaxy contains at least two trillion solar masses of material, of which only one-twentieth can be seen inside our Sun's orbit, hence our Milky Way has a massive dark matter halo.

The Solar Neutrino Problem

The nuclear reactions that light up the Sun generate huge numbers of neutrinos, which stream out of the interior in a matter of seconds, unimpeded by their interaction with matter. At any one time, trillions of these neutrinos are passing through your body every day. Physicists designed detectors to count these neutrinos but discovered that far fewer of them were present than predicted from knowledge of the temperature and reactions inside the core of the Sun. In the 1970s, this became known as 'the Solar Neutrino Problem', and was only resolved in 1998 with the discovery that neutrinos 'oscillate' from one type to another. Unless the detector was designed to interact with all three forms of neutrinos (electron, muon and tau), it cannot count them properly. Revised estimates are now in agreement with the predicted number of solar neutrinos.

Cosmologists prefer to call this *cold dark matter* (CDM) because it has to be slow-moving (cold) in order for it not to dramatically alter the amount of large-scale structure seen in the universe today such as superclusters, filaments and walls of galaxies spanning hundreds of millions of light years.

Cold dark matter

The Bullet Cluster of galaxies (1E-0657-558) also shows how dark matter can be detected indirectly. Hot gas discovered by NASA's Chandra X-Ray Space Observatory is seen as two pink clumps in the image on page 98 and contains most of the 'normal', or baryonic, matter in the two clusters. The bullet-shaped clump on the right is the hot gas from one cluster, which passed through the hot gas from the other larger cluster during the collision. An optical image from Magellan and the Hubble Space Telescope shows the galaxies in orange and white. The blue areas in this image show where astronomers find most of the mass in the clusters. The concentration of mass is determined using the effect of so-called gravitational lensing, where light from distant objects is distorted by intervening matter. Most of the matter in the clusters (blue) is clearly separate from the normal matter (pink), giving direct evidence that nearly all of the matter in the clusters is dark.

Bullet Cluster

The hot gas in each cluster was slowed by a drag force, similar to air resistance, during the collision. In contrast, the dark matter was not slowed by the impact because it does not interact directly with itself or the gas except through gravity. Therefore, during the collision the dark matter clumps from the two clusters moved ahead of the hot gas, producing the separation of the dark and normal matter seen in the image. If hot gas was the most massive component in the clusters, as proposed by alternative theories of gravity, such an effect would not be seen.

Separation of dark and normal matter

The Bullet Cluster
provides direct proof
for the existence of
dark matter.

DARK ENERGY

Independently of the CMB measurements, other teams of astronomers continued traditional studies of distant galaxies using supernovae as 'standard candles' with the goal of obtaining a better measure of the Hubble constant. Supernovae of Type-1a are created when a white dwarf in a binary star system accretes enough mass from its companion star to explode as a supernova. From studies of these events in nearby galaxies with precisely measured distances, it was discovered that these supernova rise to a maximum luminosity that is relatively constant across many different supernovae. By using this standard luminosity and comparing it to the apparent brightness of supernovae of the same class

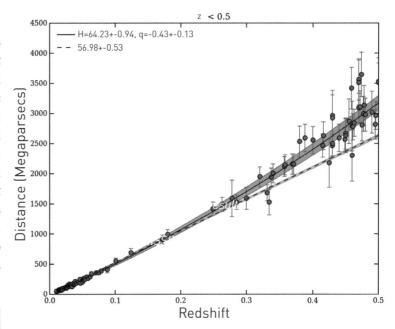

A display of supernova data from the Reiss Survey showing that expansion is accelerating.

Standard Candles

Astronomers use parallax to calculate distances to relatively nearby stars – measuring the angle and the amount of displacement six months apart, when the Earth is at the two extremes of its orbit. For distant stars and galaxies too remote for parallax, astronomers use 'standard candles' – that is, stars whose brightness is known. The most widely used are *Cepheid variables* and Type 1a supernovae. Cepheids are stars whose luminosity varies on a regular cycle. Astronomer Henrietta Swann Leavitt discovered that the period of variability is linked to its brightness. Type 1a supernovae are white dwarf stars that are stealing mass from a red dwarf companion star and explode when they reach critical mass. As the explosion always occurs at the same mass, the luminosity is always the same. If you know a star's luminosity, you can calculate its distance. Supernovae are so bright they show up in distant galaxies.

detected in distant galaxies, the distances to these remote systems can be determined.

In 1998, the High-Z Supernova Project, led by Adam Reiss, and the Supernova Cosmology Project, led by Saul Perlmutter, announced from their independent studies of several dozen supernovae that from about five billion years ago the expansion rate of the universe has not been proceeding at the constant rate predicted by Hubble's Law. What their studies showed was that it has, in fact, been accelerating as the universe has got older. Measurements from a more-distant supernova called SN 1997ff located ten billion light years away did not show signs of this acceleration, so it must have begun sometime between five and ten billion years ago.

From considerations of General Relativity, this cosmic acceleration means that there does indeed exist something like Einstein's 'cosmological constant' in the equations of Big Bang cosmology. Because this cosmological constant has the same value (density) in every cubic metre of space in the cosmos, the force that it produces increases as the volume of space increases. As the universe expands, more volume is created, and so this cosmological force increases in time and causes the accelerated expansion of the universe.

The two studies using supernovae and the CMB point to one and the same phenomena, and so Einstein was right after all, but for some reason this cosmological constant effect 'turned on' only about six billion years ago.

DARK ENERGY ▶ *a component to the contents of the universe that accounts for its accelerated expansion rate.*

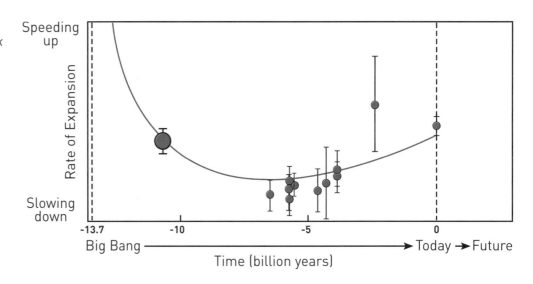

A graph showing how the current dark energy expansion is very recent. Each dot represents the measurement of an individual supernova and the implied expansion rate at its redshift.

HOW MUCH OF THE UNIVERSE IS DARK?

An alternative method of quantifying how dominant this component was in the inventory of the cosmos was to literally 'weigh' the entire cosmos using details of its expansion rate (the Hubble Constant) together with the measured properties of the cosmic microwave background (CMB). The advent of the NASA Cosmic Background Explorer (COBE) in 1989 led to high precision measurements of the CMB using the Far-Infrared Absolute Spectrophotometer (FIRAS) instrument, which obtained a temperature of 2.7260 ± 0.0013 K together with minute variations (±0.000001 K) in its distribution across the sky measured by the Differential Microwave Radiometer (DMR) instrument.

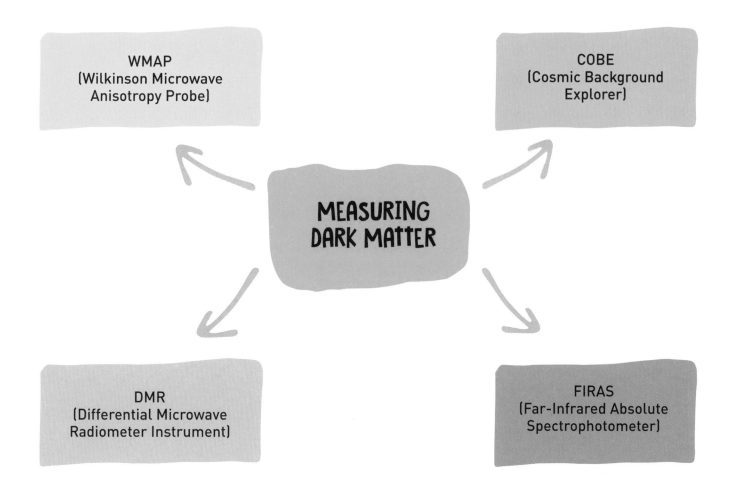

WMAP
(Wilkinson Microwave Anisotropy Probe)

COBE
(Cosmic Background Explorer)

MEASURING DARK MATTER

DMR
(Differential Microwave Radiometer Instrument)

FIRAS
(Far-Infrared Absolute Spectrophotometer)

A subsequent NASA mission called the Wilkinson Microwave Anisotropy Probe (WMAP) studied the variations that COBE's DMR instrument had detected, and was able to use it to solve for three different cosmological components simultaneously: ordinary baryonic matter, dark matter and dark energy. Baryonic matter was what astronomers had been studying for over a century in the form of luminous stars, interstellar and intergalactic gas, and degenerate objects such as white dwarfs, neutron stars and black holes. Dark matter was not likely to be ordinary matter, but might include heavy neutrinos, unknown exotic particles, WIMPS and CDM.

Dark energy was the name given to the equivalent of the cosmological constant, Λ, in the Einstein-de Sitter models. The new models were called the Λ-CDM models. What WMAP was able to deduce in observations from 2001 to 2010 was that the pattern and intensity of the cosmic microwave anisotropy (the minute bumps in the all-sky CMB light) could be fit with:

- 4.5 per cent of ordinary 'baryonic' matter,
- 22.7 per cent dark matter, *and a whopping*
- 72.8 per cent dark energy.

A follow-on study between 2009–2013 by the European Space Agency's Planck spacecraft detected even smaller irregularities in the CMB and derived improved value:

- 4.9 per cent ordinary 'baryonic' matter,
- 26.8 per cent dark matter *and*
- 68.3 per cent dark energy.

So, slightly more baryonic matter and dark matter and slightly less dark energy, but still a huge amount of energy/mass unaccounted for.

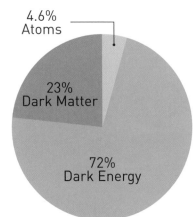

4.6%
Atoms

23%
Dark Matter

72%
Dark Energy

The percentages of dark energy, dark matter and normal matter in the cosmos according to WMAP data.

George Smoot

Born in 1945 in Yukon, Florida, he attended the Massachusetts Institute of Technology and received his bachelor's degrees in physics and mathematics in 1966 and then went on to receive his PhD in particle physics in 1970. At the Lawrence Berkeley Laboratory in California, he collaborated with Louis Alvarez to develop a differential radiometer to measure very accurately the intensity of the CMB from different regions of the sky using an instrument flown on a Lockheed U-2 jet. He succeeded in detecting for the first time the dipole effect in the CMB. He subsequently proposed that a similar instrument be flown on a satellite – a concept accepted by NASA and incorporated as the DMR instrument into the Cosmic Background Explorer launched in 1989. Along with John Mather, Smoot was awarded the 2006 Nobel Prize in physics. In addition to writing popular books and articles on cosmology, Smoot has made numerous appearances on TV programmes such as 'The Big Bang Theory' and 'Are You Smarter than a 5th Grader?' In the later instance he was only the second contestant to win the $1 million prize.

STANDARD MODEL CANDIDATE PARTICLES

What exactly is dark matter? The roughly five per cent of the universe that is in objects derived from protons, neutrons and even black holes, are all built from the quarks, so they cannot be double-counted as forms of dark matter. That leaves the peculiar particles called neutrinos.

Big Bang cosmology predicts from the primordial element abundances and the ratios of protons to neutrons that there can be at most three families of neutrinos, which have already been discovered and are called the electron, muon and tauon neutrinos. The models also predict just how many neutrinos were produced during the Big Bang. If we multiply this number by the mass of a single 'cosmological' neutrino, we get a value for the total cosmological mass and density contributed by these neutrinos.

Neutrinos

During the 1970s, it was proposed that neutrinos could contribute a considerable amount of mass to the universe to dramatically affect what kind of universe – open and infinite or closed and finite – we lived in. If the sum of the electron, muon and tauon neutrino masses were 12eV (electron volts – see page 111), then there would be enough mass in the cosmic neutrino background to actually 'close' the universe, which is not indicated by the CMB observations. A more stringent constraint comes from the observed large-scale structure in the universe. If the neutrinos are too massive, they will smear out small-scale structures such as clusters of galaxies. Models of cosmological structure formation and evolution indicate that the sum of the neutrino masses cannot be greater than a few tenths of an eV. This is also consistent with Standard Model mass limits on the three neutrino families derived from accelerator and other experiments. So, within the currently known family of elementary particles, and with neutrinos eliminated, there are no candidate particles which have been discovered so far that could explain dark matter. This leads us to the important question of what is matter, and what physical systems could account for the known properties of dark matter?

Cosmic neutrino background

PARTICLES BEYOND THE STANDARD MODEL

Many different candidate particles have been proposed over the decades to account for 'missing mass' and dark matter. However, the most promising candidates are modest extensions of the Standard Model that incorporate Supersymmetry. These are *neutralinos*, *axions*, *Kaluza-Klein* particles and *sterile neutrinos* (see pages 138–9 for more details).

REVISIONS TO GENERAL RELATIVITY

One solution to the dark matter 'problem' is that our equation for gravity used to develop Big Bang cosmology may be somehow incomplete. In 1983, physicist Mordehai 'Moti' Milgrom suggested that general relativity could be tweaked slightly in what he called Modified Newtonian Dynamics (MOND). At great distances, the force of gravity on a star orbiting a galaxy would vary not as the inverse-square but as $1/r$, where r is the distance between the star and the centre of gravity of the galaxy. This adds a new acceleration-dependent term to Newton's Law of Universal Gravitation, which is a weak–gravity limit to General Relativity. Although it can be tuned to account for the dark matter haloes of galaxies, it does not work for clusters of galaxies and larger cosmological structures. Also, MOND theory does not completely eliminate the need for dark matter at every scale, so its utility as an alternative to dark matter is disputable.

Perhaps a greater criticism than its not accounting for the known tests of General Relativity and cosmology is that extensions of General Relativity by the MOND approach do not lead to a theory that is relativistically correct for all observers. *Relativistic invariance* (*covariance*) means that all observers in the universe, regardless of their state of motion, will observe the same natural laws. This apparently is not the case for MOND theories since the additions to the equations of General Relativity which would be required to account for dark matter cannot be couched in relativistic terms. Because the modifications are acceleration-dependent, they would violate the principle of equivalency in which the gravitational and inertial masses of an object are no longer identical as required by the principle of covariance, which is the cornerstone of General Relativity.

The detection of gravitational waves in 2015–16 also eliminates many possible versions of MOND, which strongly implies that simple MOND revisions to General Relativity will not apparently solve the dark matter problem.

The Israeli physicist Mordehai Milgrom found one potential solution for the dark matter problem in so-called 'Modified Newtonian Dynamics'.

Chapter Six
WHAT IS MATTER?

The Building Blocks of the Universe – Fermions and the Anatomy of Matter – Quantum Mechanics – Leptons – Quarks – Quantum Field Theory – Electromagnetic Interaction – Werner Heisenberg – Strong Interaction – Weak Interaction – Standard Model – Symmetry Breaking – Higgs Boson

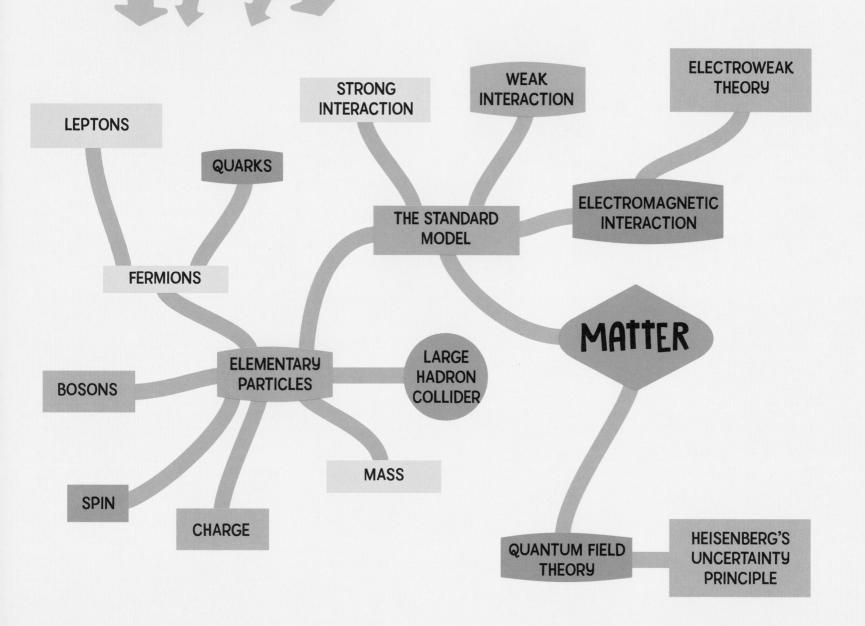

THE BUILDING BLOCKS OF THE UNIVERSE

The scope of cosmology encompasses the origin, evolution and future of all things that make up the universe. Space and time, following the relativity revolution, are included in this. But the most tangible ingredient is the matter that makes up the stars, galaxies and you, yourself. Among the deepest questions faced by modern cosmology is: how do we explain the origin of matter and its nature and interactions? For this we need to delve into what is known about atoms and their constituents; a subject called nuclear physics.

Newton was known for laws of motion and universal gravitation.

Tracks of elementary particles from collisions within 'atom smashers'.

The history of nuclear physics, and the discovery of the fundamental forces in nature, is a subject as deep and as complex as the entire pageant of events that brought astronomy from the elementary ideas of Kepler and Newton, to the enormous detail that we now have. The history of physics was greatly accelerated by the development of technologies such as 'atom smashers' that could shake loose the fundamental particles from which matter is comprised and help quantify the forces through which they act.

FERMIONS AND THE ANATOMY OF MATTER

The periodic table of the elements is a familiar listing of the fundamental elements that make up all forms of common matter. It is, basically, a catalogue of stable (and unstable) forms of matter composed of protons and neutrons, which together form the nucleus of each atom, and electrons, which surround the nucleus. For example, water is made up of one atom of oxygen and two atoms of hydrogen. The 'classical' explanation of what holds the molecule together is that each hydrogen atom has one electron in its outer shell, held there by *electromagnetic forces*. The oxygen atom has an incomplete outer shell – it lacks two electrons. The oxygen atom 'shares' the two electrons belonging to the hydrogen atoms and the *residual electromagnetic forces* involved in this interaction are what holds the water molecule together.

Until the 1960s, this was the limit of our understanding of the building blocks of all matter. Since then, physicists have discovered that hiding behind this basic atomic structure is an even more fundamental collection of elementary particles and forces. On the one hand we have a small number of particles that build up the material world. On the other, we have a separate set of particles responsible for producing the forces between the material particles.

BARYON ▶ *subatomic particle that has a mass equal to or greater than a proton. Baryons belong to the quark-based hadron family of particles.*

Particle soup

The identification of the fundamental particles in all matter was a laborious process that began in 1897 when J. J. Thompson discovered the electron, receiving the Nobel Prize in 1906 for this effort. And this process is still not complete. The last matter particle, called the *'tau neutrino'*, was discovered in 2000 at the Fermilab accelerator by a team of 54 physicists. But it is thought that there may be additional elementary particles yet to be discovered. The known elementary particles, responsible for all forms of familiar matter, have now been studied at high precision and their properties exactly determined.

J. J. Thomson discovered the electron in 1897, beginning a search for the fundamental matter particles that would not be completed until 2000.

QUANTUM MECHANICS

As we have explained, classical physics describes nature at the scale of cannon balls and galaxies. *Quantum mechanics* describes nature at the smallest scales of atoms and subatomic particles. It deals with the mathematical description of the motion and interaction of subatomic particles, incorporating the concepts of 'quantization of energy', 'wave–particle duality', the 'uncertainty principle', and the 'correspondence principle'. (These concepts are described in more detail elsewhere in this book.)

One of the great discoveries made with the introduction of quantum mechanics was that elementary particles can be described, not only in terms of their charge and mass, but by their 'spin'. This is an intrinsic property of matter that has no analogue in the non-quantum world, and very loosely represents a property of particles analogous to its namesake.

Particles' spin

QUANTUM ▶ *(plural QUANTA), discrete packets of subatomic energy; the smallest amount of energy that can be involved in an interaction.*

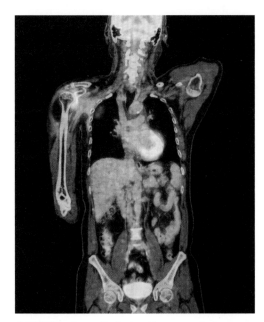

A PET (Positron Emission Tomography) scan is one example of how the knowledge of antimatter can influence modern technology.

In order to account for how atoms generate spectral lines at precise wavelengths (see spectroscopy, pages 27–8) and other phenomena of material particles, these elementary particles may only have ½ of a quantum unit of spin. This results in a variety of related phenomena. For example, only two electrons may be present in the same quantum state such that one has plus (+) ½ unit of spin and the other has minus (-) ½ unit of spin. The statistics of such particles was described by Enrico Fermi and Paul Dirac. We now call all matter particles fermions, after Fermi. In 1928 Paul Dirac predicted the existence of antimatter (see page 135). This was confirmed in 1932 by Carl Anderson's discovery of the anti-particle for the electron. It is called the *'positron'*. It has the same spin and mass as an electron, but instead of a negative electric charge, it has a positive electric charge. If a particle and its anti-particle are brought together, they immediately annihilate into a burst of energy, according to Einstein's famous $E=mc^2$. So, each fermion has its own anti-particle,

Positron

which immediately doubles the number of elementary matter particles. Fermions are divided into two families: *leptons* and *quarks*.

Particle soup
'If I could remember the names of all these particles I'd be a botanist.'
– physicist Enrico Fermi.

SPIN ▶ *a property of an elementary particle analogous to rotational movement (intrinsic angular momentum), which confers magnetic field and electric charge in macro molecules. It is not a perfect analogy because the particle does not actually spin!*

THE LEPTONS

The only familiar particle in the set of 12 elementary particles known as *leptons* is the *electron*. (An additional set of 12 anti-particles also exists.) Each electron is partnered with its own neutrino particle based on phenomena associated with radioactive decay. However, by 1995 two more generations of particles, more massive cousins to the electron, had been found: the muon and the *tauon*. As with the electron, these additional particles have their own neutrinos so that in total there are six particles in the electron family, called for convenience *leptons* (meaning 'light'). *Neutrinos* are bizarre particles that interact with matter only very weakly. They can zip through a light year of matter and not get absorbed. At first they were thought to carry no mass at all, but since the detection of 'neutrino oscillations' in the Japanese Super-Kamiokande Detector in 1998, it is now recognized that neutrinos carry a very small mass of less than one electron volt for all three masses combined.

LEPTON ▶ *elementary particle that contains one unit of electric charge and responds only to the electromagnetic force, gravitational force and weak force – not the strong force. They include electrons, muons, tauons, neutrinos and their antimatter variants.*

Electron-Volts

Particles can be described in terms of their mass in kilograms. But these numbers are cumbersome, so physicists prefer to use $E=mc^2$ to derive a mass unit of E/c^2 where E is measured in electron-volts. To further simplify mass units, they also take $c = 1$ and refer to particle masses in millions of volts (MeV) or billions of volts (GeV):

- An electron has a mass of about 0.5 MeV;

- A muon is over 200 times as massive at 105 MeV;

- A tauon is 3600 times as massive at 1.8 GeV.

THE QUARKS

A second family of fermions, called *quarks*, is even more exotic. Physicists are not very imaginative when it comes to naming new particles, unlike astronomers who carefully consider many possibilities when they name planets. Murray Gell-Mann originally proposed the nonsense name 'quork'. Then he came across a line in James Joyce's *Finnegans's Wake* – 'Three Quarks for Muster Mark' and adopted the spelling.

In 1964, the quark model for nuclear matter was independently proposed by the American and Russian-American physicists Murray Gell-Mann and George Zweig. The basic idea was that protons, neutrons and other massive particles known by 1960 are not fundamental, but are composed of still smaller particles called Up (U) and Down (D) quarks. By combining the Up and Down quarks in the right way, you can build up the proton (UUD) and the neutron (DDU) and other massive particles that had been discovered at particle accelerator labs, such as the neutral π meson (U and anti-U). As more and more massive particles were discovered, Strange (S), Charmed (C), Top (T) and Bottom (B) had to be added to this family. The last of these, the Bottom quark, was discovered in 1995.

Combinations of quarks

QUARK ▶ *elementary particle; quarks combine to form composite particles such as hadrons. There are six types: Up, Down, Top, Bottom, Strange and Charmed, plus their anti-matter variants.*

Quark	Symbol	Spin	Charge	Baryon Number	S	C	B	T	Mass*
Up	U	½	+⅔	⅓	0	0	0	0	1.7–3.3 MeV
Down	D	½	−⅓	⅓	0	0	0	0	4.1–5.8 MeV
Charm	C	½	+⅔	⅓	0	+1	0	0	1270 MeV
Strange	S	½	−⅓	⅓	−1	0	0	0	101 MeV
Top	T	½	+⅔	⅓	0	0	0	+1	172 GeV
Bottom	B	½	−⅓	⅓	0	0	−1	0	4.19 GeV(MS) 4.67 GeV(1S)

HADRON ▶ *subatomic particle, such as a baryon or a meson, that can take part in the strong interaction.*

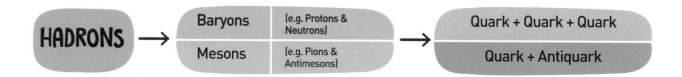

MESON ▶ *subatomic particle, intermediate in mass between an electron and neutron, composed of one quark and one anti-quark.*

As of 2018, this collection of six quarks and six leptons (together with their 12 anti-particles) is believed to be complete, and forms the basis for describing states of matter that can be generated at collision energies up to at least 13 TeV. Just as the periodic table of the elements accounts for all chemical and nuclear physics states of matter, this table of 24 fundamental fermions covers the products of all known astrophysical phenomena involving energy production within stars via thermonuclear fusion, and even explains why supernovae explode and create dense remnants.

A theory of matter that only describes the properties of fermions does not also explain how these particles interact in space and time. These interactions are caused by forces, but only three non-gravitational forces are needed to describe all possible interactions. Separate from, but intertwined with, the elementary fermions is a separate set of particles that are responsible for these forces, but to explain how they work requires delving into a powerful theory of forces and matter called *quantum field theory*.

BOSONS			force carriers spin = 0, 1, 2, …		
Unified Electroweak spin -1			Strong (color) spin =1		
Name	Mass GeV/c^2	Electric charge	Name	Mass GeV/c^2	Electric charge
γ photon	0	0	g gluon	0	0
W–	80.4	-1			
W+	80.4	+1			
Z^0	91.187	0			

QUANTUM FIELD THEORY

Each elementary fermion is surrounded by a field. For electrons it is called the *electric field*. For quarks it is called the *gluon field*. But these fields actually consist of innumerable particles that, according to Heisenberg's Uncertainty Principle, cannot be directly detected or observed. They are called virtual particles because they only 'virtually' exist. The exchange of these individual *quanta* of the fermion fields gives rise to the three elementary forces. All of these force-mediation particles carry exactly one unit of quantum spin. The statistics of particles with integer units of quantum spin was worked out by the Indian theoretical physicist Satyendra Nath Bose and Albert Einstein and called Bose-Einstein statistics. Particles with integer quantum spin (0, 1, 2…) are simply called 'bosons'.

BOSON ▶ *subatomic particle such as a photon with zero or integral spin.*

Richard Feynman helped to found the field of quantum electrodynamics.

Richard Feynman
This American physicist was one of the founders of quantum electrodynamics along with Julian Schwinger and Shin'ichiro Tomonaga in the late 1940s. He developed many mathematical techniques for the rapid calculation of the probabilities of interactions between electrons and photons, including a diagrammatic representation of these many processes (called Feynman diagrams) and a sum-over-histories formulation for adding up all of the possible quantum outcomes of an interaction. His philosophical outlook on the search for a 'Theory of Everything' led him into many arguments with developers of string theory, accusing them of promoting fantasy speculations and turning them away from concrete experimental science. Before his death in 1988, he was instrumental in uncovering the cause of NASA's Space Shuttle Challenger accident in 1986.

THE ELECTROMAGNETIC INTERACTION

The electromagnetic force, or interaction, is transmitted by virtual photons. Particles surrounded by the virtual photons of the electric field feel the electrostatic force when these virtual photons are exchanged. The exchanges are called virtual processes, and they can be very complicated. The simplest is the exchange of a single virtual photon. The next-most-complex is when the emitted virtual photon suddenly produces an electron–positron pair, which then decays back into a virtual photon. The mathematical techniques of *quantum electrodynamics* (QED), developed in the late 1940s, describe these processes in terms of Feynman diagrams. These also provide a diagrammatic way to represent the elementary ingredients of virtual processes, although they are not intended to be a 'photograph' of what is actually occurring. In fact, physicists do not know what elementary particles look like or if it even makes sense to try to describe them in terms of any 'visual' property.

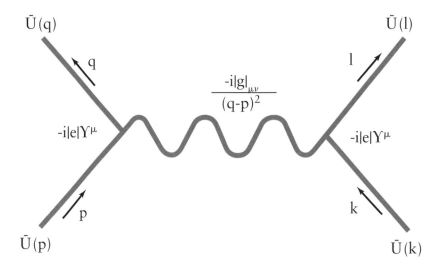

A Feynman diagram for electrons interacting by the exchange of a single photon

QUANTUM ELECTRODYNAMICS ▶ *field theory that unites quantum mechanics and Special Relativity to explain how light and matter interact.*

Heisenberg's Uncertainty Principle

This principle, devised by the German physicist Werner Heisenberg, states that it is not possible to know both the position and velocity of an object at the same time. It is expressed by the following formula:

$$\Delta E \Delta T \geqslant \frac{h}{4\pi}$$

This formula says that there is a limit in quantum mechanics to how well we may simultaneously know how much energy (ΔE) exists in a quantum process if it is measured for a specific duration of time (ΔT) The limit is set by the value of *Planck's Constant* $h = 4.1 \times 10^{-15}$ eV s. This means that some types of phenomena can occur for which the conservation of energy seems to be violated, but only if they occur for less than a certain amount of time. For example, the mass of an electron is 510,000 eV, so for the virtual process where an electron and a positron occur, the total energy of the pair is 1.2 million eV.

The relationship says that: $\Delta T \geqslant 2.7 \times 10^{-22}$ s.

This is interpreted to mean that even in empty space, an electron-positron pair can temporarily appear so long as they last no more than 0.00000000000000000000027 seconds (2.7×10^{21} s) before vanishing. Virtual photons carry energy and when they are emitted by one fermion they have to be absorbed by a second fermion before this time limit is reached. In quantum field theory, this process is the cause of the electromagnetic interaction, with the virtual photon as the mediator of the force.

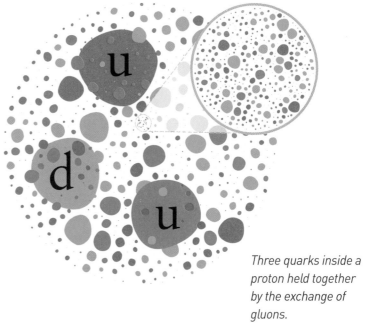

Three quarks inside a proton held together by the exchange of gluons.

WERNER HEISENBERG

Werner Heisenberg was a German physicist who was one of the developers of modern quantum mechanics. His work on matrix mechanics in 1925 was one of the key papers that described the behaviour of electrons within atoms at a time when the atomic structure of matter was still a murky subject of study. His research led to the discovery of a limit to how well one can simultaneously measure what are called conjugate variables such as momentum and position, or time and energy, referred to as the Heisenberg Uncertainty Principle. His matrix mechanics represented the outcomes of any measurement of an atomic system as a set of numbers that expressed the probability of a transition between one state and another. It also suggested that during a transition, electrons simply did not exist until the moment of the change itself, at which point they acquired measurable properties.

The Strong Interaction
Figure-eight combinations of gluon-mediated forces.

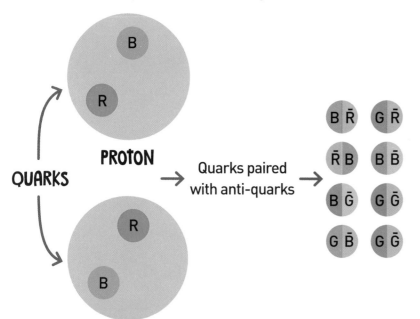

QUARKS

PROTON

NEUTRON

Quarks paired with anti-quarks →

BR̄ GR̄
R̄B BB̄
BḠ GḠ
GB̄ GḠ

There are only eight gluons because the quark model is based on a symmetry called SU (3). There is no ninth gluon required, which in any case would be colourless and not tkae part in the strong interaction.

THE STRONG INTERACTION

Quarks have to be confined inside protons and neutrons. This arises from the fact that quarks exchange particles called *gluons*. Like photons, gluons also carry exactly 1 unit of quantum spin. However, because quarks possess a new kind of charge called *'colour'*, these three charges (arbitrarily called *red*, *blue* and *green*) have to be reflected in the colour charges of the gluons, leading to eight possibilities, such as red + anti-blue, red + anti-green etc. in exactly eight combinations. (This gluon-mediated force is what was previously called the π meson [or pion] nuclear interaction in the years before the quark model was proposed. It was thought that the exchange of pions was the carrier of the strong interaction.) With the advent of the quark model, pions were seen as composite particles made up of pairs of quarks and their anti-quarks, and the strong force carrier is now identified as the exchange of the more elementary gluons.

The Strong Interaction
The Strong interaction binds together to form protons and neutrons.

q(u) + q(u) + q(d) = Proton (uud)

q(u) + q(d) + q(d) = Neutron (uud)

STRONG FORCE (OR INTERACTION) ▶ *one of four fundamental forces governing all matter; it binds quarks together to form composite particles such as baryons (e.g. protons and neutrons).*

Electromagnetic v Strong Interactions

Particle	Electrical charge	Colour Charge	Interacts with
electron	-e	–	Photons
electron neutrino	0	–	–
up quark	+⅔	red, green, blue	photons, gluons
down quark	+⅓	red, green, blue	photos, gluons
photon	0		–
gluon	0	colour + anticolour	gluons

THE WEAK INTERACTION

Some particles are known to spontaneously decay in time. The neutron, for example, decays into a proton (+ charge) and electron (- charge) and an electron anti-neutrino in about 881 seconds. This decay process is described in terms of a third force called the *weak nuclear force*. Unlike the electromagnetic force with a single massless photon, or the strong force with eight massless gluons to mediate their effects, the weak force is mediated by three heavy particles called the W⁺, W⁻ and Z⁰ *intermediate vector bosons*. Like their cousins, they also have a spin of 1 unit, but they are very massive. The W⁺ and W⁻ are each 80 GeV while the Z⁰ is about 91 GeV; masses that are about equal to the entire nucleus of the element Strontium. Because of their massiveness, these vector bosons have exceedingly short ranges, which is why the weak nuclear force is much weaker than the strong nuclear force.

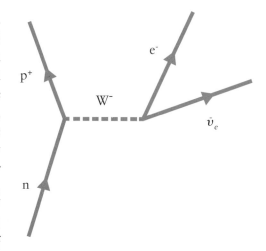

A neutron decay mediated by the weak interaction provided by the W⁻ boson.

> **WEAK INTERACTION** ▶ *one of four fundamental forces governing all matter; it works at short distances between subatomic particles and causes radioactive decay. In weak reactions, particles may disappear or reappear.*
>
> *β⁻- decay: neutron* ▶ *proton + electron + electron antineutrino.*
>
> *β⁺- decay: proton* ▶ *neutron + positron + electron neutrino.*

THE STANDARD MODEL

Quantum electrodynamics

The collection of 12 elementary fermions and 12 elementary bosons (1 photon, 3 vector bosons and 8 gluons) makes up the fundamental collection of matter particles and their force mediators. An entire universe of mathematical techniques has evolved since the 1940s to accurately describe their precise interactions for a wide variety of physical processes. For the electromagnetic force we use the tools provided by *quantum electrodynamics* (QED). For the strong interactions we use the tools of *quantum chromodynamics* (QCD). At about the time that the final touches were being applied to a detailed theory of the weak interaction in the late-1950 and early-1960s, a very different concept entered theoretical physics called *'symmetry breaking'*.

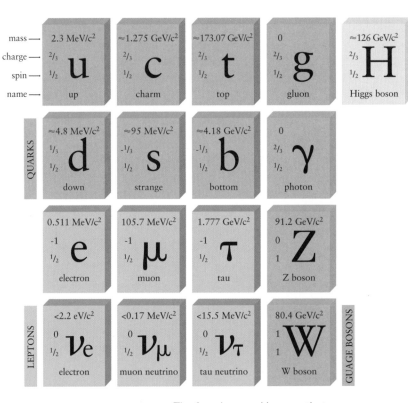

The fermions and bosons that make up the Standard Model.

QUANTUM CHROMODYNAMICS ▶ *a quantum field theory in which the strong interaction is described in terms of an interaction between quarks mediated by gluons. Quarks and gluons are both assigned a quantum number called 'colour'.*

SYMMETRY BREAKING

Physicists had previously been aware of the role played by various symmetries in nature. In 1915, the German physicist Emmy Noether discovered this relationship in order to solve a difficult problem in general relativity. Simply put, if a process looks the same when it is shifted in time, this means that energy is conserved. If it looks the same when it is shifted in space, that means that momentum is conserved.

Emmy Noether
Born in Erlangen, Germany in 1882, Noether received her doctorate in mathematics in 1907 at the University of Erlangen, where as a women she worked for seven years without pay. Then in 1915 she was invited by Felix Klein and David Hilbert to join the mathematics department at the University of Gottingen. Between 1919 and 1933, she was recognized for her brilliance in algebra and developed the concepts of rings and fields, and went on to develop the concept that conservation principles in physics are related to underlying symmetries, which came to be known as the Noether Theorem. This insight was the basis for most of the modern work in theoretical physics during the 1940s and beyond.

Emmy Noether uncovered the relationship between conservation and symmetries.

What the British theoretical physicist Peter Higgs, and independently his Belgian contemporaries Francois Englert and Robert Brout, discovered in 1964 was that the symmetries in the mathematical theories of which the strong, weak and electromagnetic forces are an example, could be broken by introducing a new particle (or field). Physicists Steven Weinberg, Sheldon Glashow and Abdus Salam ran with this idea in the late 1960s and devised a theory that combined both the mathematics of the electromagnetic and weak interactions and symmetry-breaking, for which they received the Nobel Prize in 1979.

Peter Higgs was awarded the Nobel Prize for Physics in 2013.

The *'electroweak theory'* proposes that the electromagnetic and weak interactions are all mediated by massless bosons (photon, W^+, W^- and Z^0). However, even the fermions (electrons and quarks) are massless as well. Under these conditions there is an exact symmetry between how these bosons interact. This means that the electromagnetic and weak forces look the same in terms of their strengths.

Electroweak theory

HIGGS BOSON

There is also a new particle called the *'Higgs boson'*, with a quantum spin of zero, which also interacts with the weak and electromagnetic force carriers along with the matter particles themselves. The Higgs boson at very high energies starts out being massless, so at high energies the mass of the fermions and bosons remains zero and their interactions preserve symmetry. However, as the interaction energy is lowered, the Higgs boson gains mass due to interactions with itself. As it gains mass, so do the fermions and bosons. The particles that most strongly interact with the Higgs bosons (the W and Z^0 bosons) gain the most mass, while weakly interacting particles gain very little (neutrons) or no mass at all (photons, gluons). This mechanism for breaking symmetries in the descriptions of particle interactions is called spontaneous symmetry breaking (SSB), and this particular process is called the *'Higgs Mechanism'*.

Higgs bosons

The Higgs Mechanism

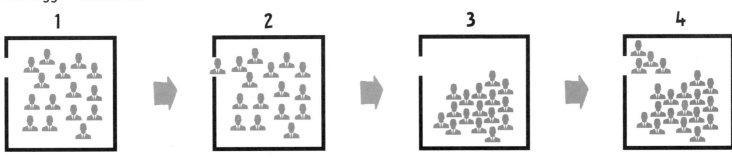

1. *The Higgs field can be thought of as a molasses-like viscosity to empty space. Imagine, for example, a room full of people.*

2. *A movie star enters the room and shakes hands and signs autographs, but this slows down her motion.*

3. *She finds it harder to move freely, as though she has suddenly gained weight, or the people in the room act like molasses to retard her motion.*

4. *A less popular actor enters the room and attracts fewer admirers, so feels less viscosity and 'weight gain'.*

Stephen Weinberg developed the electroweak theory along with Sheldon Glashow and Abdus Salam.

It was not until 2012 that the Higgs boson was actually detected at the Large Hadron Collider at CERN. The Standard Model, with its 24 fermions and bosons and its 25th particle – the neutral Higgs boson – is now considered complete. It is based around a fixed core of exactly 25 fundamental fermions and bosons together with two mathematical systems for making the necessary calculations: quantum chromodynamics and electroweak theory. It has proved to be exceptionally accurate in making predictions for all known interactions at energies up to the highest operating limits of the Large Hadron Collider by 2018.

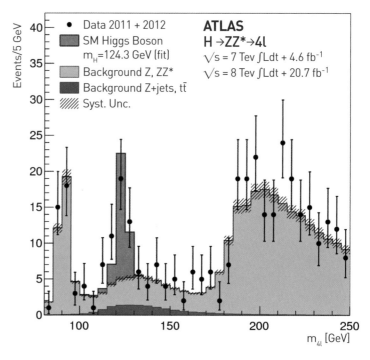

Data from the LHC showing the detection of the Higgs boson.

The ATLAS detector within the Large Hadron Collider.

The Large Hadron Collider

The Large Hadron Collider (LHC) is the world's largest and most powerful particle accelerator. It was built between 1998 and 2008 near Geneva, Switzerland, by the European Organization for Nuclear Research (CERN) and a collaboration of over 100 countries at a cost of $4 billion. To describe this vast 27-km (16-mile) machine would be an essay in superlatives. Within its magnetically confined ring system, protons circulate in opposite directions and collide at several specific points along the circumference. At these points, massive detector systems track and catalogue all of the collisions and their products for later recall. Over one billion collisions are recorded every second, and require networks of supercomputers to keep up with the data-gathering. The highest collision energy achieved by 2018 is 13 trillion electron-Volts (13 TeV). The collider requires $1 billion each year to pay for the electricity needed to power it and vast quantities of liquid helium to operate it. By some estimates, the Higgs boson, discovered in 2012, cost $13 billion to discover.

TESTING THE STANDARD MODEL

The powerful mathematical techniques describing the properties of the forces and particles, together with supercomputers and advanced particle accelerators such as the Large Hadron Collider and Fermilab, allow the Standard Model to make dozens of fundamental predictions of specific types of interactions and particle decays. These can be tested to high precision.

The number of possible particles that result by simply combining the six quarks and six anti-quarks in patterns of twos (mesons) is exactly 39 mesons. Of these, only 26 have been detected as of 2017. For the still heavier three-quark baryons, the quark patterns predict 75 baryons containing combinations of all six quarks. There are 31 of these predicted baryons which have not been detected so far. These include the lightest missing particles, the *Double Charmed Xi* (U, C, C) and the *Bottom Sigma* (U, D, B), and the most massive particles, called the *Charmed Double-Bottom Omega* (C, B, B) and the Triple-Bottom Omega (B, B, B). To make life even more interesting for the Standard Model, other combinations of more than three quarks, called '*exotic baryons*', are also possible.

Glueballs are one of the most novel, and key, predictions of the Standard Model, so not surprisingly there has been a decades-long search for these waifs among the trillions of other particles that are also routinely created in modern particle accelerator labs around the world. Glueballs are not expected to live very long. They carry no electrical charge and so are perfectly neutral particles. From the various theoretical considerations, there are 15 basic glueball types that differ in what physicists term '*parity*' and angular momentum. By 2015, the f-zero(1500) and f-zero(1710) had become the prime glueball candidates, with the f-zero(1710) the best candidate consistent with experimental measurements and its predicted mass.

So the Standard Model, and the six-quark model it contains, makes specific predictions for new baryon and meson states to be discovered. All told, there are 44 ordinary baryons and mesons that remain to be discovered!

The particle accelerator at Fermilab near Chicago, Illinois is used to accurately test predictions about how particles interact.

125

Chapter Seven

BEYOND THE STANDARD MODEL

Theory of the Unified Field – Symmetry in Mathematics – Symmetry in Physics – Grand Unification Theory – Supersymmetry – Matter and Antimatter – Λb^0 and its Antiparticle – The Particle Desert – Particles Beyond the Standard Model

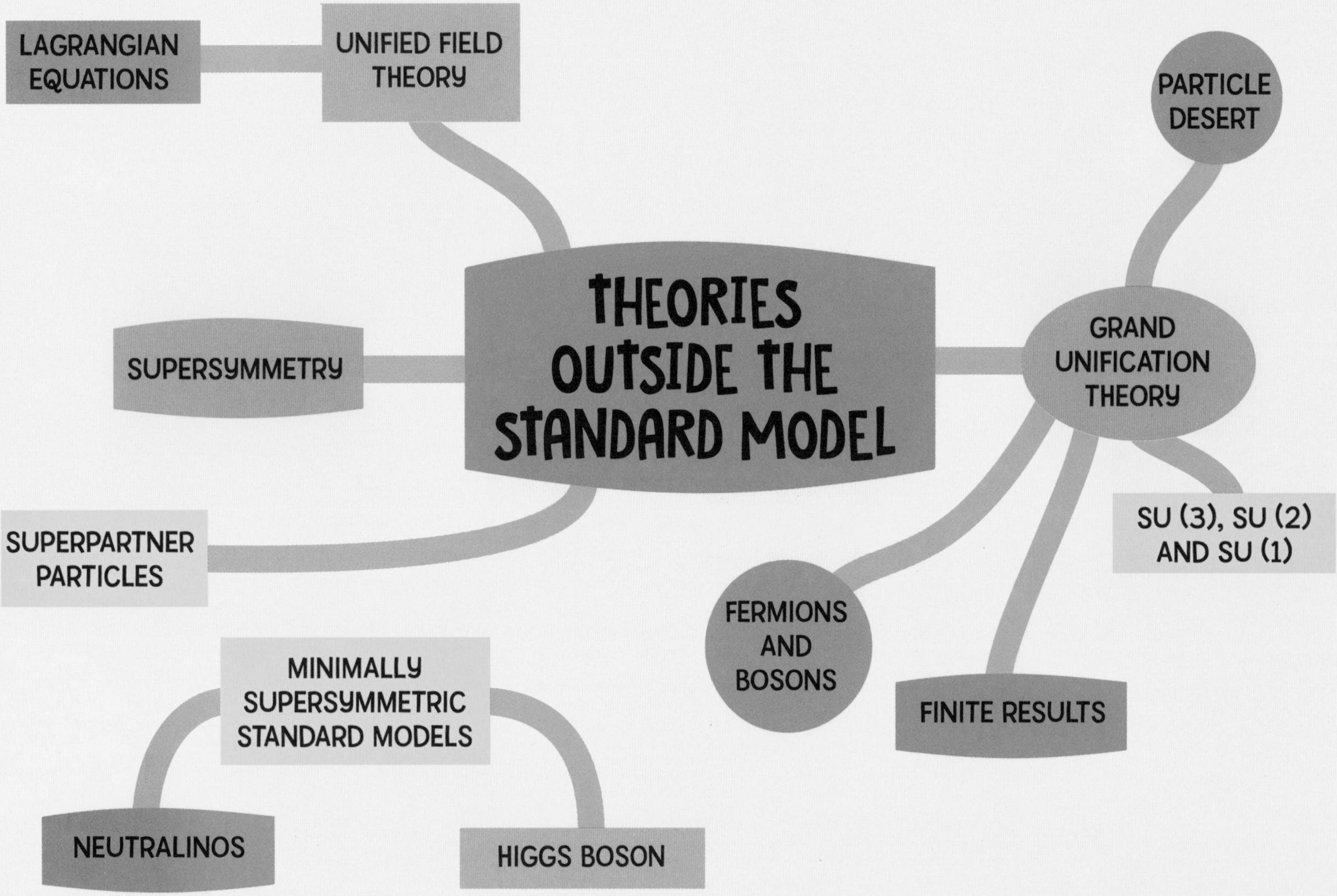
THEORIES OUTSIDE THE STANDARD MODEL
LAGRANGIAN EQUATIONS
UNIFIED FIELD THEORY
SUPERSYMMETRY
SUPERPARTNER PARTICLES
MINIMALLY SUPERSYMMETRIC STANDARD MODELS
NEUTRALINOS
HIGGS BOSON
PARTICLE DESERT
GRAND UNIFICATION THEORY
SU (3), SU (2) AND SU (1)
FERMIONS AND BOSONS
FINITE RESULTS

THE THEORY OF THE UNIFIED FIELD

By the time Einstein was refining his 1915 Theory of General Relativity, it was already obvious to him and to many other physicists that there were two great forces operating in the universe: gravity and the electromagnetic force. Both had exceedingly long, if not infinite, ranges and were described in detail by two very different mathematical descriptions. On the one hand, Maxwell's equations and their relativistic extensions provided by Special Relativity described electromagnetic fields and forces. On the other, Einstein's new Theory of General Relativity described gravity as a different kind of field. Einstein spent most of his life trying to find a single mathematical formulation, a *'Unified Field Theory'*, in which gravity and electromagnetism could be reinterpreted as aspects of a new field of nature. Einstein never favoured the tools of quantum mechanics, and so these efforts led nowhere. Meanwhile, the mathematical language of Quantum Field Theory, which was used to describe the strong, weak and electromagnetic forces, steadily evolved into a rich arena for at least unifying these three non-gravitational forces. These productive ideas involved the concept of symmetry as the cornerstone idea, which facilitated finding mathematical similarities among the forces.

SYMMETRY IN MATHEMATICS

Take an ordinary featureless cube and rotate it in the horizontal plane by 90 degrees. Each face looks the same. This is called rotational invariance or rotational symmetry. There are exactly 24 ways that you can rotate a cube so that after the operation the cube looks the same. Because these rotations are additive and always result in another operation that is already a member of the set of 24 rotations, this is called a 'group of rotations'. We say that the symmetry of the cube is invariant under the group of 24 operations, which is called by mathematicians the O_h Octahedral symmetry group. In crystallography, the complex but regular shapes of mineral crystals can be classified into exactly 32 possible symmetry groups.

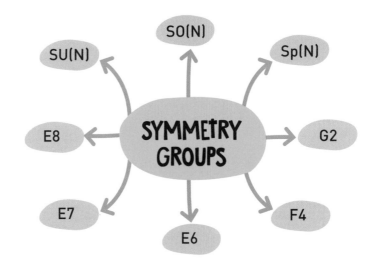

The mathematician Eli Cartan long ago systematically classified all of the possible 'simple' symmetry groups into eight categories: SU(N), SO(N), Sp(N), G2, F4, E6, E7 and E8, the smallest of which, called the Special Unitary Group 2 or SU(2), had 1 symmetry operation, while E8 had 248.

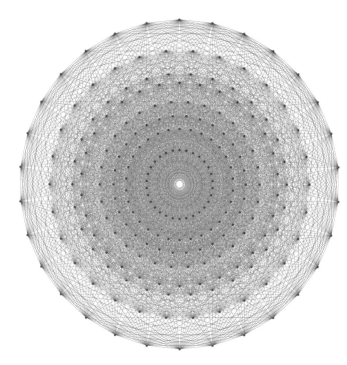

A diagram relating the various symmetry operations within the E8 group.

SYMMETRY IN PHYSICS

In physics, the properties of the interactions between particles amount to writing a complex equation called the Lagrangian, which describes all of the ways that the particles and fields interact among themselves. A simple way to think of this is an equation that is the difference between the kinetic and potential energies of an interacting system, written in mathematical terms. The first term is the square of the field's momentum change in time, much like kinetic energy is the square of a particle's velocity. The second term is the potential energy of the field due to its interaction with itself and with other fields in the Standard Model.

If in the Lagrangian equation you can change the variable that

The Lagrangian

In school physics, we learned that the total energy of a closed system on the surface of the Earth is the sum of its kinetic energy, $\frac{1}{2} mv^2$, and its gravitational potential energy, $V(h) = mgh$, expressed as the formula:

$$E = \frac{1}{2} mv^2 - V(h)$$

In Quantum Field Theory, described in the previous chapter, this energy equation is replaced by a more complicated equation that, for the simplest massless scalar field represented by the variable ϕ looks like this:

$$L = -\frac{1}{2} \partial_\mu \phi^* \partial_\nu \phi \eta^{\mu\nu} - V(\phi^* \phi)$$

More complicated Lagrangians with more terms have to be considered in order to fully describe the known matter and force fields, one for each type, but nevertheless they have to be added in such a way as to preserve the known symmetries among these forces and particles. The search for a unified field theory or *Grand Unification Theory* (GUT) largely involves discovering what new terms have to be added to the system's Lagrangian and in what combinations.

SCALAR ▶ *a quantity in physics that has magnitude only, such as 10 kg, 20 cm etc, and no other characteristic.*

SCALAR FIELD ▶ *a scalar quantity applied to every point in a given space, such as the background temperature of the universe.*

SYMMETRY ▶ *an aspect of a system that stays the same after some transformation. Knowing that some parts stay the same facilitates the discovery of what quantities are conserved in the system such as energy, momentum, charge, etc.*

represents time from +t to -t and still get the same identical physical process; this means that energy is being conserved and the Lagrangian is symmetric in time. Likewise, if you can displace the physical process in space by changing the space variable +x to -x, and the new process is indistinguishable from the former process, the process is said to conserve momentum. Emmy Noether in 1919 discovered this intimate relationship between symmetry and conserved quantities. Noether's Theorem has been called one of the most important mathematical theorems ever proved in guiding the development of modern physics and specifically how to add terms to the Lagrangian that preserve physical symmetries in nature. You can't just jumble a bunch of fields together in an equation because the resulting mess can violate many well-known conservation theorems.

In mathematical terms, the three non-gravity forces obey the following symmetries:

- Electromagnetism – Unitary Group U(1).
- Weak interaction – Special Unitary group SU(2)
- Strong interaction – Special Unitary group SU(3).

The SU(2) group of operations has $2^2-1 = 3$ elementary operations.

The SU(3) group of operations has $3^2-1 = 8$ elementary operations.

These 8+3 operations can be related to the eight gluons and three intermediate vector bosons, which are now called the 'gauge' bosons of the symmetry groups. For example, in the strong interaction based on the three colour charges (red, green and blue), in order that the Lagrangian for the quarks obeys SU(3) symmetry in terms of the eight possible colour operations, there has to be a new field (terms) added to the Lagrangian which preserves this symmetry. This is the field provided by the eight physical gluons. But, like Russian matryoshka dolls, smaller symmetry groups could be nested in larger ones that 'unified' them.

GRAND UNIFICATION THEORY (GUT)

The concept of symmetry groups completely took over theoretical physics in the 1970s, and turned out to be a fiendishly complex language to master by other scientists, including those in related fields such as astrophysics. Nevertheless, during this decade the search for the Mother of All Symmetries that would unify the three forces has been enormously productive. There were several ground rules that had to be followed:

A unified theory of these forces would have to include the already known symmetries that were exploited in crafting the successful theories for the electromagnetic force, U(1), the weak force, SU(2) and the strong force, SU(3).

The new symmetry group would have to include the known particle types within the specific families of fermions and bosons which are observed. Any additional particles would have to meet stringent experimental tests across the energy ranges already explored. The new group would have to be mathematically well-behaved when it came to making calculations about whether certain processes would occur. The calculations had to lead to finite results.

The process of finding a symmetry group large enough to:
- accommodate the Standard Model;
- reproduce the electroweak symmetry breaking at low energy;
- yield a single interaction strength at high energy; *and, at the same time,*
- account for the known families of particles,

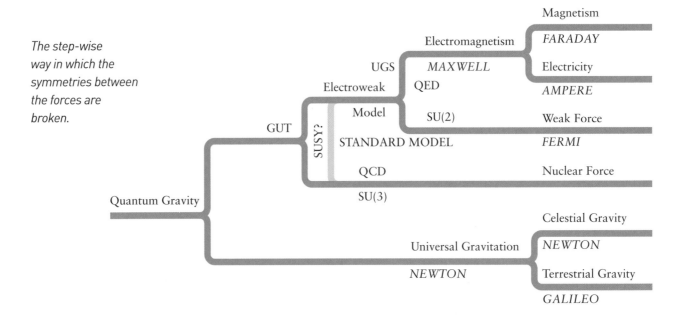

The step-wise way in which the symmetries between the forces are broken.

Finding the right Lagrangian was often a challenging, tedious and – more often than not – 'trial and error' exercise. The only symmetry groups that appeared to have the correct attributes were identified by the labels SU(5), SO(10), E6 and E8. These initial attempts to unify the three forces using Group Theory methods were very exciting, beginning with the work on SU(5) by Howard Georgi and Sheldon Glashow at Harvard University in 1974. Although it has fallen out of fashion since 1980, SU(5) exposed physicists, and in particular the new generation of graduate students, to some of the basic principles of how unification would proceed via the study of symmetry groups.

As for the electroweak symmetry broken by the Higgs boson, in SU(5) this even larger symmetry would be broken by a new super-massive Higgs-like field. The symmetry group required $5^2-1 = 24$ massless bosons that mediated this new symmetry. In counting the number of these fields, SU(3) for the strong interaction provided $3^2-1 = 8$ of these bosons, the weak force provided $2^2-1 = 3$ more and finally U(1) for electromagnetism provided 1 boson for a total of 12 bosons that were known. That means that there were 12 additional fundamental bosons required to make SU(5) symmetry work. Amazingly, when this SU(5) symmetry is broken by the new family of supermassive Higgs bosons, the 12 new 'X and Y' bosons gain an enormous mass near 10^{15} GeV – some 100 trillion times the mass of the proton. They do this because of their interactions with the supermassive Higgs boson. This is completely analogous to the electroweak Higgs boson causing the W and Z bosons of the weak force to gain their mass, causing the weak force to look different from the electromagnetic force at much lower temperatures.

Over the next decade, many other symmetry-breaking schemes were investigated in order to unify the strong and electroweak forces. A vast number of technical difficulties have been encountered, specifically in guaranteeing that calculations would lead to finite results. Some of these problems were solved, at least temporarily, by proposing that space had up to 26 dimensions, of which only four were our current spacetime of near-infinite extent. The other dimensions were wrapped up and smaller than 10^{-30} cm.

Another feature of these early unification theories was the implication of the symmetry-breaking mechanism that made the strong and electroweak forces distinct at lower energy. There would have to be a vast 'desert' between 10^{15} GeV and about 300 GeV where no new stable particles would ever be found. This was a chilling announcement for experimenters who had always been rewarded with new particles each time they built newer, more expensive, particle accelerators. For Big Bang cosmology, this was actually good news because across that vast energy range and timescale, you would not

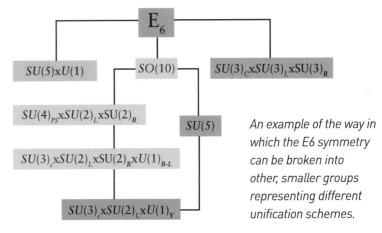

An example of the way in which the E6 symmetry can be broken into other, smaller groups representing different unification schemes.

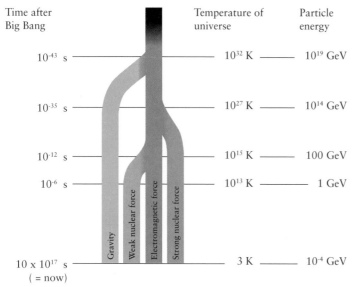

Time after Big Bang

10^{-43} s

10^{-35} s

10^{-12} s

10^{-6} s

10×10^{17} s (= now)

Gravity

Weak nuclear force

Electromagnetic force

Strong nuclear force

Temperature of universe

10^{32} K ——— 10^{19} GeV

10^{27} K ——— 10^{14} GeV

10^{15} K ——— 100 GeV

10^{13} K ——— 1 GeV

3 K ——— 10^{-4} GeV

Particle energy

The way in which the forces become unified in a simple unification scheme.

have to worry about new particles coming and going. The Standard Model particles were all that you needed.

Although there is no single Grand Unification Theory that is universally agreed upon, the contenders do have several common and significant predictions to make about the physical world. Two critical temperatures emerge, signifying the onset of two major symmetry-breaking phases. Much as ice turns to water at 0°C (32°F) and to steam at 100°C (212°F), the laws governing matter and its interactions change abruptly at the temperatures characterizing the electroweak and GUT unification energies. The energies at which these 'crystallizations' occur are truly fantastic. The electroweak transition is predicted to occur at 200–300 GeV, which is reachable by modern accelerators such as the Large Hadron Collider, whereas the GUT transition requires 1000 trillion GeV, which is not likely ever to be reachable by human technology.

For these theories to predict unification, they will also require the existence of new families of particles. For example, SU(5) requires 24 gauge bosons, of which the Standard Model bosons represent only 12. There would have to be another 12 'supermassive' bosons with masses near the GUT scale of 10^{15} GeV.

The EW theory correctly predicted the existence of the W^+, W^- and the Z^0, and also a new particle called the Higgs boson. Rather than the Higgs boson being a family of two particles, in GUT theory, a 25-member family of even more massive Higgs bosons seems to be needed in some of the simple versions. Also, many of the GUT candidates also predict that the proton should eventually decay. Is there any experimental evidence that the strong, weak and electromagnetic forces actually become unified at a single energy as the interaction energies are increased? The answer seems to be a guarded 'yes'.

At very low energies, the strength of the electromagnetic interaction is determined by the so-called fine structure

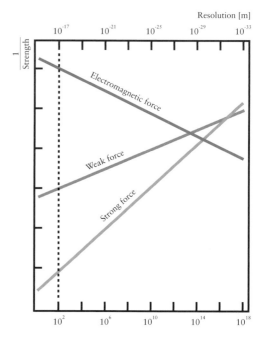

Resolution [m]

10^{-17} 10^{-21} 10^{-25} 10^{-29} 10^{-33}

$\frac{1}{\text{Strength}}$

Electromagnetic force

Weak force

Strong force

10^2 10^6 10^{10} 10^{14} 10^{18}

Grand Unification Theory predicts that the forces will unify in strength at very high energies.

constant, $\alpha = 0.00730$. At energies near 90 GeV, however, experiments show that its value has increased to $\alpha = 0.00782$. The corresponding strong interaction constant measured at 34 GeV has a value of $\alpha = 0.134$ and at 90 GeV its value has decreased to $\alpha = 0.119$. These changes with increasing energy at least show that the electromagnetic interaction is getting more like the strong interaction, and the strong interaction is weakening. All GUTs predict this behaviour. Nevertheless, it is difficult to extrapolate measured trends at energies below 100 GeV and argue that by 10^{15} GeV they are consistent with a single interaction strength.

FINE STRUCTURE CONSTANT ▶ *a number (close to $1/_{137}$) that relates to the strength of the electromagnetic force and governs how elementary charged particles (electrons and muons) and light (photons) interact.*

SUPERSYMMETRY

During the 1980s, a completely new way of unifying particles and forces was developed thanks to the discovery in 1978 of a possible new symmetry in nature called Supersymmetry, discovered by Julius Weiss and Bruno Zumino.

Supersymmetry

The Standard Model consists of fermions with spin-$1/2$ and bosons with spin-1. These fields are described in the Lagrangian as separate terms and are not unified due to their differing spin values. Supersymmetry (called SUSY) is a new symmetry of nature that allows fermions and bosons to mathematically change into each other. This is because each fermion and each boson is paired with a new particle. For example, the photons with spin-1 are paired with new particles called *photinos* with spin-$1/2$. In the Lagrangian equation, you have to include these new particles, not only to insure that Supersymmetry is maintained, but that the calculations for various processes and interactions lead to finite answers. The masses of these SUSY particles can be upwards of 1 Tev. However, the Large Hadron Collider has so far not detected any collision that indicates these particles exist, or can be generated in the expected manner.

This new symmetry was able, mathematically, to transform fermions into bosons. This meant that there was now a mathematical scheme for unifying the force carriers (bosons) with the particles they act upon (fermions). What Supersymmetry also did was to propose that an entirely new collection of particles exists. Each normal particle such as quarks, electrons, gluons and photons had superpartner particles called *squarks*, *selectrons*, *gluinos* and *photinos*. These particles, with

Supersymmetry requires additional particles beyond the Standard Model.

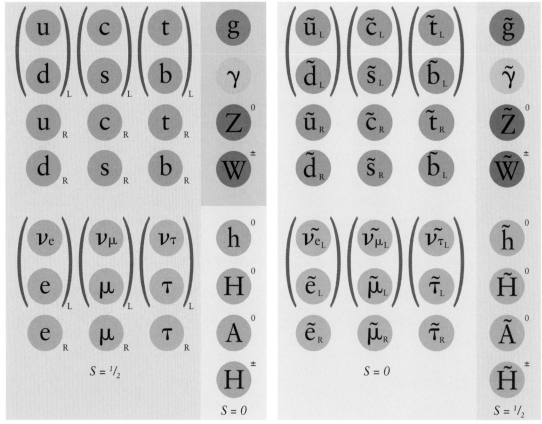

Existing particles   SUSY particles (MSSM model)

masses above 1 TeV or more, were far more massive than the ordinary Standard Model particles.

What was also interesting, and very exciting, was that when the Supersymmetry transformation was twice applied to a particle to convert it from a normal fermion to a SUSY boson and then back to a normal fermion, it also changed the particle's location in spacetime. This kind of spacetime translation explicitly involved the metric $g_{\mu\nu}$. This meant that gravity via its spacetime metric symbolism was automatically being brought into the theory. For this, Supersymmetry was heralded as an already unified theory of gravity and matter. This led to investigations of the mathematics of Supersymmetric gravity, or 'supergravity', along with other applications of Supersymmetry in the late 1970s and 1980s. Various versions of Standard Models extended by SUSY were also developed. The search was now on for aspects of the Standard Model that indicated new physics consistent with *Minimal Supersymmetric Standard Models* (MSSM) that incorporated the new particles and symmetries.

SUPERSYMMETRY ▶ *the principle that proposes a relationship between fermions and bosons and that each particle in one group has a paired 'superpartner' in another group. It is intended to fill in gaps and inconsistences in the Standard Model.*

Estimating the Grand Unification Energy Scale

According to the MSSM model with Supersymmetry, the strong interaction is found to change with energy according to $y = 5 - 1.1 \times$ and the electromagnetic force changes according to $y = 63 - 2.4 \times$, where y is the inverse-strength of the force and $\times$ is the logarithm of the energy (e.g. $\times = 5$ means 10^5 GeV). For what energy, $\times$, will the forces have equal strength? Set the equations equal to each other $(63 - 2.4x = 5 + 1.1 \times)$ and solve for $\times$ to get $\times = 17$ so at an energy of about 10^{17} GeV the forces should be equal.

MATTER AND ANTIMATTER

In the mid-1920s, the basic rules of quantum mechanics and the behaviour of electrons were discovered and rendered into a mathematically precise theory suitable for making calculations and predictions for a wide variety of atomic phenomena, especially in spectroscopy. But the theory did not satisfy Einstein's theory of relativity. Physicist Paul Dirac in 1928 almost over-night developed a relativistic quantum theory for the electron, but in order to preserve certain symmetries in the mathematics he had to propose a partner particle to the electron which was identical to the ordinary electron except that it had a positive and not a negative electric charge. This 'anti-electron', now known as a *positron*, was soon discovered by Carl David Anderson in 1932. From then on, the idea of 'antimatter' has found its way into numerous theories of physics and a multitude of science fiction stories and has found practical applications in technology, such as positron emission tomography (PET) scanning.

A photograph of electrons spiralling to the right and antielectrons spiralling to the left as they move through a fixed magnetic field.

The Λb° particle

Antimatter

An electron is an elementary particle with -1.6×10^{-19} coulombs of electric charge, a quantum spin of ½ Planck's unit of spin, and a mass of 9.1×10^{-31} kg. What Dirac discovered in the mathematics is that to make the relativistic equations have the correct symmetries, you need to also have a partner particle with $+1.6 \times 10^{-19}$ coulombs of charge, a quantum spin of ½ Planck's unit of spin, and a mass of 9.1×10^{-31} kg. In other words, it would look exactly like an electron but with a positive charge. When the electron and the anti-electron are combined mathematically, they lead to a quantum state with the properties of the vacuum, but in which the $E=mc^2$ energy of the combined particles appears as a pair of photons. In other words, combining a particle and its antiparticle liberates $E=2mc^2$ of energy. Other particles such as the neutron are their own antiparticles because they carry no charge. Quarks can also have antiparticles because if a quark carries $+\frac{1}{3}$ electric charge, its antiquark will carry $-\frac{1}{3}$ electric charge. They may not annihilate if their 'colour charges' are different, however. Once again the end state has to have the quantum properties of the vacuum with no net electric or colour charge in the final state.

Λb° AND ITS ANTIPARTICLE

One of the biggest cosmological challenges has been to explain why matter dominates the stars and galaxies in the visible universe, and why there is not an equal amount of antimatter. For decades it has been known that some nuclear reactions are, in fact, not symmetric between matter and antimatter. Most recently, the decay of the exotic particle called Λb° and its antiparticle shows a statistically significant difference favouring matter in some 6,000 measurements made at the Large Hadron Collider and announced in 2017. The Λb° consists of three quarks – Bottom, Up and Down – and has a decay lifetime of 1.5 picoseconds, which leads to more matter than antimatter in the final decay products involving leptons. There are, however, no Standard Model processes that lead to more baryons (quarks) than antibaryons (antiquarks) that are stable states. To make more matter than antimatter, some new process or symmetry has to be broken, and this is one of the key issues driving the search for physics beyond the Standard Model, and the hope that such new physics will at last account for why we are here as creatures of matter.

THE PARTICLE DESERT

In the history of all previous colliders beginning in the 1950s, something new has always been found to move the development of our explanatory capabilities forward. In the 1970s it was the discovery of quarks. In the 1980s it was the W and Z^0 particles, and as recently as 2012 it was the Higgs boson. All these particles were found below an energy of 200 GeV. But now, as the Large Hadron Collider spent 2017 operating at 13,000 GeV, a desperate mood has set in. No new particles or forces have been discovered in this new energy landscape, even though the theoretical decay steps for massive superparticles has been well-developed, mathematically.

There may exist a vast 'desert' with no new particles to be discovered.

Most of the theories that go beyond the Standard Model provide ways to unify the strong force with the electromagnetic and weak forces. The predicted energy where this unification happens is about 1,000 trillion GeV. According to popular Supersymmetry theory calculations, there should be a large population of new particles above an energy of 1,000 GeV. Each of these would be a partner to the 25 known Standard Model particles, but far more massive. Some of these particles, such as the neutralino, are even candidates for dark matter, discussed in a

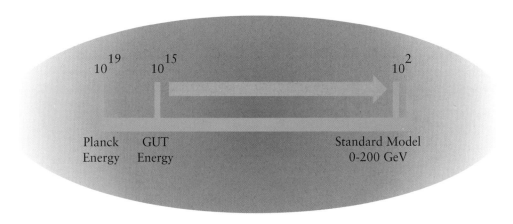

later chapter (see page 177). Above the masses of these new Supersymmetry particles, however, there ought to be no new particles to discover from perhaps 100,000 GeV to the GUT energy of 1,000 trillion GeV. Without this particle desert, any particles in this mass range would be able to cause the proton to decay much faster than current limits predict. Although this Particle Desert may serve an important theoretical need to stabilize the proton against decay, it is a devastating prediction for experimental physics. It cost $13 billion to find only a single new particle, the Higgs boson, at the Large Hadron Collider. How do you find funds to build larger colliders to explore the even vaster and more costly Particle Desert?

So, although it is a frustrating prospect that no new particles may exist in this desert, it may actually be a vital feature of our physical world that literally prevents all matter from disintegrating! Unlike the Sahara Desert, which we can at least drive through to get to more fertile regions beyond, there are apparently no easily reachable oases of new particles to which physicists can target new generations of expensive colliders.

PARTICLES BEYOND THE STANDARD MODEL

NEUTRALINOS

Supersymmetry requires that there exist partner particles to the 25 known particles. However, their masses are enormous. The most promising of these are the cousins to the ordinary *neutrinos* called neutralinos. There are many different versions of these Minimally Supersymmetric Standard Models (MSSMs) and they differ in the exact numbers of new particles they introduce, depending on the various assumptions of the models. Generally, the lightest neutralino, called the N_1^0, is stable and only interacts with matter via the weak force and gravity. This makes it a perfect candidate for the hypothetical weakly interacting massive particles (WIMPs) and cold dark matter (CDM) needed to account for cosmological dark matter, specifically in the way it causes clustering and large-scale structure formation across the cosmos.

AXIONS

Symmetries are important in physics because they relate to conserved quantities. For example, if you swapped -t for +t in equations describing the energy of a particle, you would get the same energy value. This is a reflection of the conservation of energy related to time-inversion symmetry. In nuclear physics, if you examined an interaction between particles by changing the sign of the charges (C), and reflecting the interaction in a mirror (parity, P), the interaction should look the same. In other words, replace a particle by its antiparticle and reverse the interaction left to right. The resulting interaction should be the same. This is called CP invariance. This CP invariance is obeyed by nearly all particle interactions except for the decay of a particle called the K-meson (Kaon) through the Weak Interaction. However, it is obeyed by the strong interaction. To account for this difference, a new particle called the *axion* has been proposed. It would be even less massive than a neutrino with a mass expected to be between 10 micro-eV and 1000 micro-eV. With a mass of about 5 micro-eV, it would be able to account for dark matter.

KALUZA–KLEIN PARTICLES

In the 1920s, Oscar Klein found a way to unify the gravitational and electromagnetic forces by adding a fifth dimension to spacetime. This would increase the number of *space* dimensions from three to four, except this additional dimension would be finite in size, and vastly smaller than atomic scales. Further investigations of this five-dimensional theory of general relativity also showed that all of the normal particles would be the lowest rungs of ladders in which each additional rung represented a partner particle with progressively higher masses. The lightest of these new particles have masses between 500 GeV and 1 TeV. If they were abundant cosmologically, their interactions with normal matter soon after the Big Bang would have dramatically changed such things as the expansion rate of the universe, and the ratio of primordial helium to hydrogen.

GRAVITINOS

In the 1970s, physicists proposed the idea of Supersymmetry as a way to unify the fundamental fermions and bosons in the Standard Model. As a consequence of this new symmetry of nature, every particle in the Standard Model would be partnered with a new particle, including the hypothetical particle called the *graviton*. Supersymmetry applied to unifying gravity with the other Standard Model forces is called *'Supergravity Theory'*. Gravitinos carry a quantum spin of $\frac{3}{2}$ as a fermion partner to the spin-2 graviton particle in Supergravity Theory. They are believed to have masses below 100 GeV, and can decay into photons and neutrinos, or into a Zo boson and a neutrino. The decay times are expected to be very long, even compared to the age of the universe. These decays into photons are expected to show up as an irregularity in the spectrum of the cosmic microwave background radiation. An interesting cosmological source for gravitinos as dark matter candidates is in the decay of the particles believed to be responsible for inflation (see chapter 11), which occurred soon after the Big Bang.

STERILE NEUTRINOS

These are particles similar to ordinary neutrinos, except that they only interact via their gravity and not through the weak interaction (and gravity) that characterizes Standard Model neutrinos. There is no known constraint on the number of types of sterile neutrinos. They arise, theoretically, because Standard Model neutrinos are classified as 'left-handed' and so by symmetry there could be an equivalent population of 'right-handed' neutrinos. These would be sterile and not interact through the 'left-handed' weak Interaction. Because they can get mixed with ordinary neutrinos, searches explore situations in which neutrinos are generated. Recently, at the Fermilab accelerator in Illinois, USA, a team of researchers detected an excess of electron neutrinos produced from the muon neutrinos via the neutrino oscillation process, which could be an indirect indication of sterile neutrinos. If these neutrinos had a mass of about 7,000 eV, they could account for dark matter and could produce additional X-ray light at an energy of 3,500 keV. That, so far, has not been conclusively detected.

THE SEARCH FOR NEW PHYSICS

Teams of physicists working in the 'data dumps' of billion-dollar colliders have sifted through terabytes of information to refine the accuracy of the Standard Model and compare its predictions with the real world. The predictions always seemed to match reality and push the testing of the Standard Model to still higher energies. But at the Large Hadron Collider at CERN, among the trillions of interactions studied up to energies of 13,000 GeV, no new physics has been seen in the furthest decimal points of the Standard Model predictions since 2012.

Today, we have one such elegant contender for extending the Standard Model that involves Supersymmetry. Many versions of Minimally Supersymmetric Standard Models (MSSMs) have been mathematically developed, but most agree that starting at a mass of about 1,000 times that of a proton (1 TeV), you'd start to see the lightest of these particles as 'low-hanging fruit'.

For the last seven years of LHC operation, using a variety of techniques and sophisticated detectors, absolutely no sign of Supersymmetry has yet to be conclusively found. Searches for *squarks* and *gluinos* (the Supersymmetric partners to quarks and gluons) have turned up nothing below a mass of 2 TeV. There is no evidence for exotic Supersymmetric matter at masses below 6 TeV, and no heavy partner to the W-boson has been found below 5 TeV. Even more disconcerting, Supersymmetric particles should also appear in the virtual processes that lead to the calculations of certain important physical constants. But the high-precision values of these constants agree with calculations that do not involve these Supersymmetric processes. MSSM also gives astronomers a tidy way to explain dark matter and closes the book on what it is likely to be. Unfortunately, the LHC has found no evidence for light-weight neutralinos at their expected MSSM mass ranges.

Nature seems to favour simple theories over complex ones, so are the current theories really the simplest ones?

The 60-inch (152-cm) cyclotron at Berkeley, California, which began operation in 1939, was one of the world's first atom smashers. Today, efforts continue to learn more about the fundamental matter of the universe with the Large Hadron Collider at CERN.

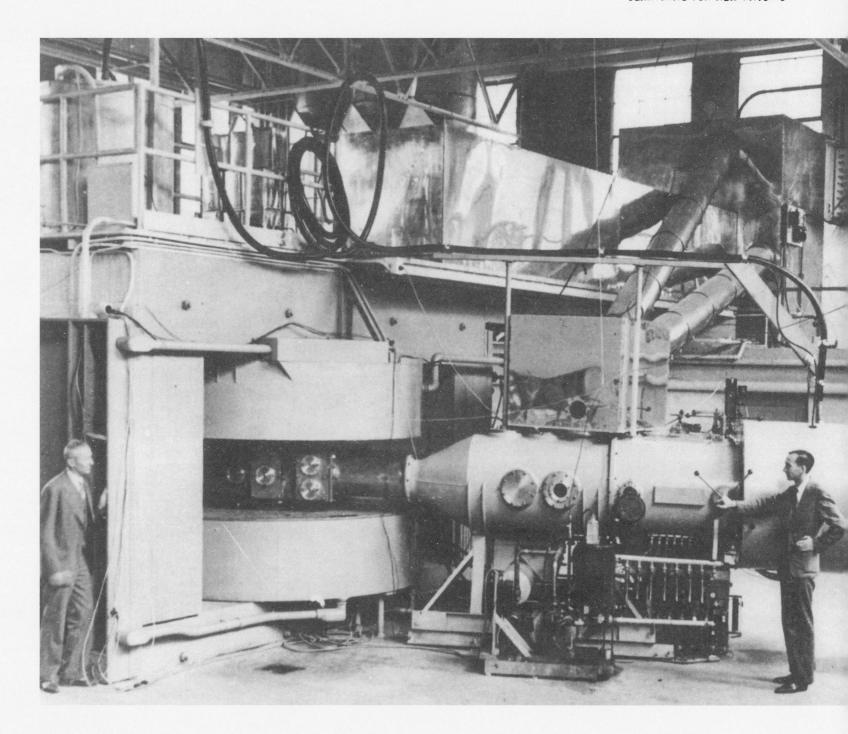

Chapter Eight
THE BEWILDERING GALAXY ZOO

The Dawn of Galactic Studies – Radio Galaxies – Quasars – Active Galaxies – Black Holes – Accretion Disks – Supermassive Black Holes

QUASARS

ACTIVE GALAXIES

NORMAL GALAXIES

HUBBLE CLASSIFICATION

SUPERMASSIVE BLACK HOLES

IRREGULARS

ELLIPTICALS

RADIO GALAXIES

SPIRALS

EXTREME STAR FORMATION

BL LACERTIDS

STARBURST GALAXIES

SEYFERTS

BLASARS

BSOS

IR GALAXIES

PROTO-GALAXIES

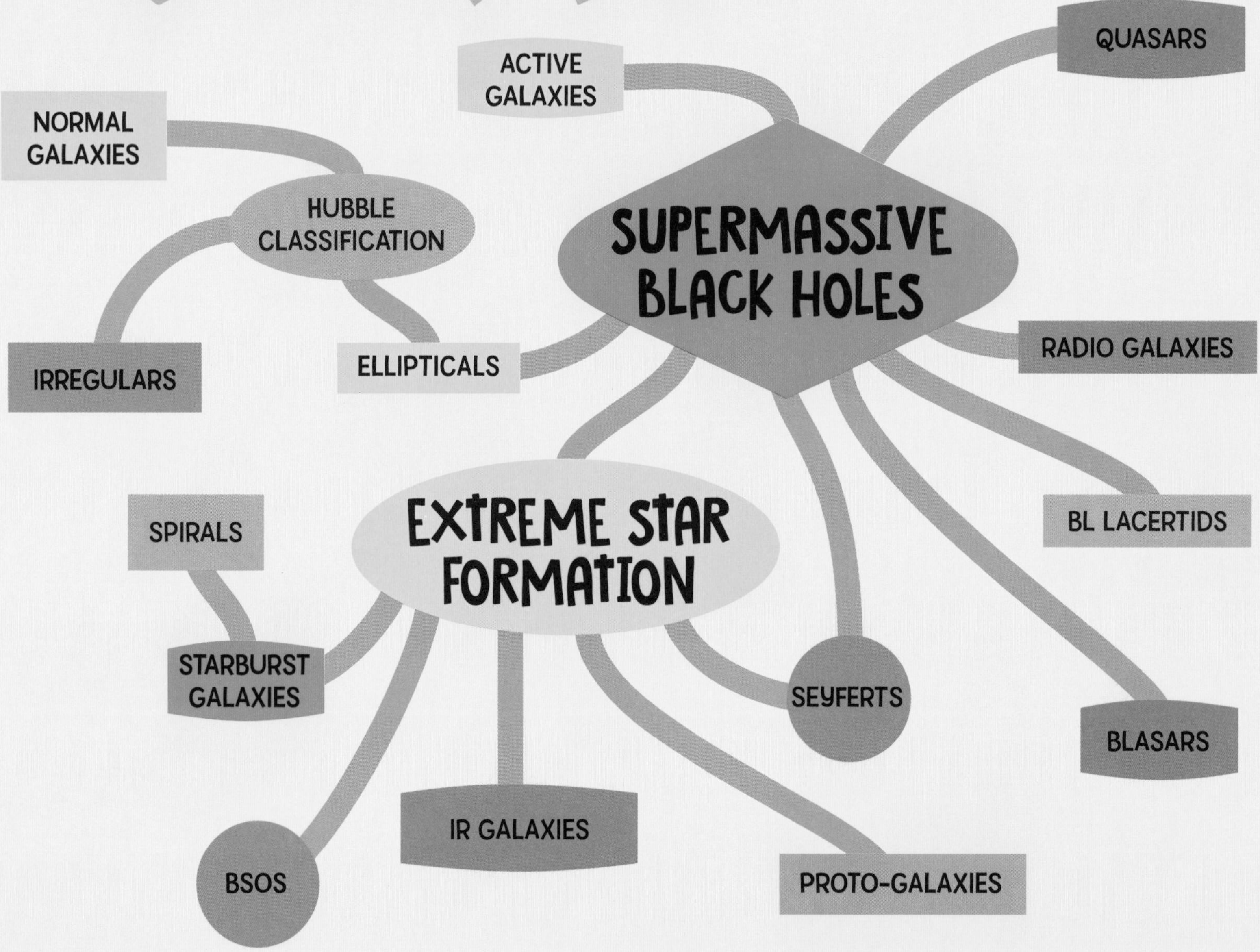

THE DAWN OF GALACTIC STUDIES

The variety of galactic forms, from ellipticals and spirals to irregular systems, seen at different viewing angles and perspectives, is astronomically bewildering, as Sir William Herschel discovered in the late 18th and early 19th centuries. By the 1930s, telescopes and photography had amassed a large catalogue of galactic forms. These were soon brought into a classification scheme by Edwin Hubble in 1936, purely on the basis of the progression of shapes he saw. Some astronomers even proposed, though mistakenly, that these forms represented the progressive evolution of galaxies from one type (elliptical) to the next (spirals and barred spirals). The advent of new ways to observe the universe through infrared, radio and x-ray telescopes also revolutionized our understanding of galactic morphology during the second-half of the 20th century.

By the late 1930s, nebulae had been classified into the four major classes: elliptical, spiral, barred-spiral and irregular. This was still the dawn of galactic studies where only a small number of galaxies had had their distances determined. Only very rudimentary information about galactic forms could be gleaned from photographs of a small number of close-by systems. Even the term 'galaxy' had not been applied to these objects and the terms 'island universe' or 'extra-galactic nebula' were still common.

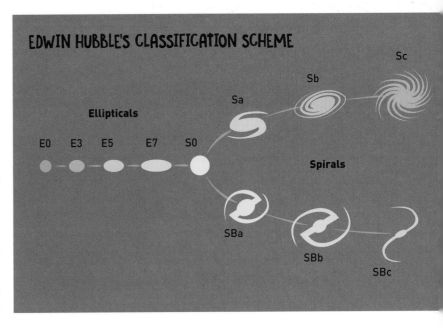

Spectroscopic techniques had only just reached the point where these faint nebulae could be detected with enough light to discern their chemistry. Their spectra were identified as similar to the light from innumerable solar-type stars that remain unresolvable even through the most powerful telescopes. The 1921 pioneering Doppler work by Vesto Slipher involving 40 of these objects revealed enormous velocities; the directions were predominantly away from the observer, especially for objects with the largest speeds. NGC584 was among the fastest-moving galaxies at that time, at +1,800 km/s. In his words, '*Certain forms of the generalized theory of relativity...indicate that very distant luminous bodies should appear to have large velocities of recession...and it may ultimately turn out that these observations give us a clue to the extent of space itself.*'

An early classification scheme for galaxy forms.

Stepping forward to 1955, dazzling photographs of many galaxies were now routinely available with enough detail to resolve stars in the closest objects such as M 33 and M 101 as well as the famous Andromeda Galaxy (M 31) and the Magellanic Clouds. Improved photographic techniques vastly increased the number of known galaxies, and by some estimates based upon

Modern photograph of M 101 the Pinwheel Galaxy – a typical spiral galaxy.

small regions of the sky, more than 75 million galaxies were surmised to exist. There are now estimated to be between 100 and 200 billion galaxies in the universe, based on a photograph called the Hubble eXtreme Deep Field (XDF).

By the mid-20th century, distance methods involving Cepheid stars and the Hubble Law relationship had found galaxies at distances of 700 million light years, in the case of the Hydra cluster, and with recessional speeds of 38,000 km/s. Big Bang expansion was now firmly being described as an explanation for these breathtaking recession speeds. It is known from Doppler studies that some galaxies such as the spirals rotate; they contain numerous stars like our Sun; and their detailed forms are to some extent determined by the location of obscuring interstellar clouds. But beyond this, little is known about their dynamics or evolution.

Hubble eXtreme Deep Field

By combining images obtained by the Hubble Space Telescope of a small region in the Fornax constellation (a patch of sky less than a tenth that of the full Moon), over a period of ten years, astronomers have identified 5,500 galaxies. This combined photograph, known as the Hubble eXtreme Deep Field (XDP) includes images of galaxies so distant we see them when the universe was less than 5 per cent of its current age. This region represents just one portion in 30 million of the whole sky. Based on this image, astronomers estimate the universe as a whole may contain as many as 100–200 billion galaxies.

RADIO GALAXIES

The detection of radio emission from the Milky Way by American pioneer radio astronomer Grote Reber in 1944 opened up a new window to the universe. Since then, many new and more sensitive 'radio telescopes' have been constructed. In addition to the radio emission from the Milky Way, which could now be mapped in great detail, numerous 'radio stars' were being discovered. Many of these were not merely points of radio light in the sky, but could be resolved and mapped using radio interferometers in which two or more radio telescopes are combined across continents to detect structure and form.

One of the most powerful extra-galactic radio sources, Cygnus A (also called 3C 405), is a double radio source located some 600 million light years distant, in which the dumbbell-shaped pair of 'lobes' are separated by about 500,000 light years. Moreover, from photographic searches with the Mount Palomar 508-cm (200-in) telescope, the centre of this radio source was found to coincide with a distorted pair of colliding galaxies.

In a growing number of cases, optical candidates could be found for these radio sources in which only a single large radio source appeared offset from the optical object. In several cases, such as Virgo A (M 87), an optical 'jet' of light could be seen emanating from the galaxy's nucleus in the direction of one of the radio-emitting lobes. Over time and at high resolution, individual plasma clouds (called plasmons) many light years across could be seen travelling down the jet as though being ejected from some unseeable source at the base of the jet and in the core of the host galaxy. The speeds of these plasmons were often measured to be large fractions of the speed of light, making them among the fastest physical phenomena seen in the universe.

QUASARS

In 1963, as optical searches for radio sources continued, one object called 3C48 was found by astronomer Allan Sandage to have nothing but a faint blue star-like object at its centre. Astronomers Jesse Greenstein and Thomas Matthews were able to obtain a spectrum for this object and discovered that its lines made no sense. The same year, Maarten Schmidt and John Beverley Oke detected the optical counterpart to 3C273 and their spectroscopic work indicated a 'redshift' of z=0.16, which means a recession speed of 16 per cent of the speed of light. It was Schmidt who correctly interpreted the wavelength shift as normal atomic lines displaced to longer wavelengths due to cosmic expansion. It was now also possible to understand the earlier spectrum of 3C48 if the spectrum wavelengths were shifted to the red by about 37 per cent, implying a recession speed of nearly 110,000 km/s. The term 'quasar' was coined by astronomer Hong-Yee Chiu in May 1964 in an article appearing in the magazine *Physics Today*.

During the 1960s, the hunt was on for more quasars, leading to a catalogue of about 40 examples by 1968. Today, more than 200,000 quasars are known – most identified from the Sloan Digital Sky Survey. Most observed quasar spectra have redshifts between z = 0.056 and 7.085. Applying Hubble's law and general relativity to these redshifts, it can be shown that they are between

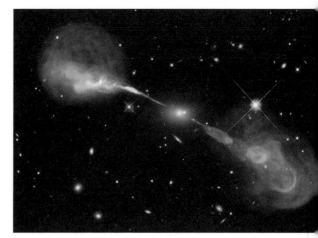

The Hercules A radio source images at optical (HST) and radio (VLA) wavelengths.

QUASARS ▶ *or 'quasi stellar radio sources', are massive sources of radio energy found in the nucleus of remote galaxies. They may contain massive black holes.*

Quasar energy production

A quasar typically produces about 10^{47} ergs/s of power. Over the course of a year this amounts to 3×10^{54} ergs/year. From $E=mc^2$, the amount of mass that needs to be annihilated to provide this energy is $3\times10^{54}/(3\times10^{10})^2 = 3\times10^{33}$ g/year. Since our Sun has a mass of 2×10^{33} g, this is about 1.5 solar masses each year. Rotating black holes can only convert about 42 per cent of rest mass into energy. So this means that about three solar masses each year in the form of stars, gas and other matter, have to be absorbed by a supermassive black hole to account for the quasar luminosity if the underlying supermassive black hole is rotating. This is generally believed to be the case from theoretical, supercomputer studies of black hole merger events.

ERG ▶ *amount of energy expended by a force of one dyne (unit of force) acting over a distance of 1 cm.*

Sloan Digital Sky Survey

The Sloan Digital Sky Survey is a mainly spectroscopic survey made at visible, infrared and ultraviolet wavelengths, which has produced detailed 3-D images of 35 per cent of the sky. It has calculated the red shift of quasars and luminous red galaxies, among other celestial phenomena. It uses a 2.5-m (8-ft) wide-angle optical telescope based at Apache Point, New Mexico.

600 million and 28 billion light years away (in terms of co-moving distance). Because of the great distances to the farthest quasars and the finite velocity of light, they and their surrounding space appear as they existed in the very early history of the universe. The most distant known quasar, J1342+0928, is at a redshift of z=7.54 and existed when the universe was only 700 million years old. We are seeing the light from this object when the first stars and galaxies were forming.

By plotting the number of quasars at each redshift, astronomers have identified an Era of Quasar Formation that occurred between redshifts of z = 0.5 and 3.0, corresponding to a period about two to five billion years ago. Today, the formation mechanism for quasars apparently is not as effective as it once was. So, fewer examples exist in our part of the universe. In fact, 3C273 with a redshift of z = 0.16 remains the closest known quasar at a distance of 2.4 billion light years. Its luminosity amounts to over four trillion stars like our own Sun. Even so, it is not the most luminous quasar known. The quasar SDSS J0100+2802 discovered in 2015 at a redshift of z = 6.3 produces 430 trillion times the light energy of our Sun, and we see its light when the universe was only 900 million years old.

Hubble telescope view of a quasar, showing interacting galaxies.

ACTIVE GALAXIES

Since the 1960s a bewildering ensemble of peculiar galaxies have been discovered. Many show indications of activity in their dense nuclei. Studies of these *active galaxies* at a variety of wavelengths from the radio and infrared to X-rays reveal three separate types of activity.

STARBURST GALAXIES

show indications of large numbers of massive stars being formed in a short span of time, and also evidence of many supernova events as some of these massive stars end their lives.

The starburst galaxy Messier 94.

SEYFERT GALAXIES

are powerful, compact sources of radio and infrared radiation usually found in the cores of spiral-type galaxies. Their nuclei often contain ionized gas travelling at thousands of km/s as though expanding from some central source that has ejected these clouds.

The Circinus-A seyfert galaxy showing complex star formation and gas ejection.

BL LACERTIDS AND BLAZARS

are galaxies with bright star-like cores that vary in optical and radio brightness over the course of months or years. The first galaxy of this type was actually misidentified as an ordinary variable star in the Milky Way and designated *BL Lacertae*. Blazars are even more variable on times as short as hours, and also produce gamma rays.

An artist's impression of a blazar.

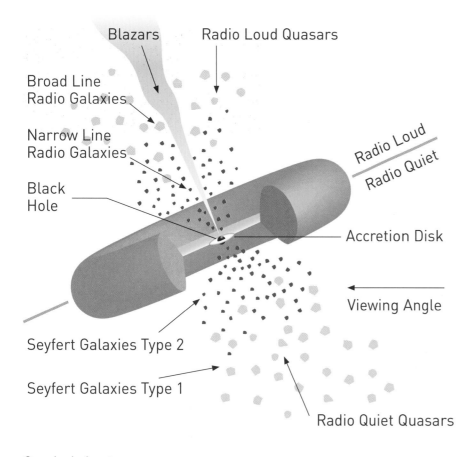

Blazars

Radio Loud Quasars

Broad Line
Radio Galaxies

Narrow Line
Radio Galaxies

Black
Hole

Radio Loud

Radio Quiet

Accretion Disk

Viewing Angle

Seyfert Galaxies Type 2

Seyfert Galaxies Type 1

Radio Quiet Quasars

Studies of active galaxies have found that many are associated with colliding galaxies in which violent collisions between interstellar clouds provide the stimulus for forming numerous massive stars. For other active galaxies such as the Seyferts and BL Lacertids, it is thought that they represent a common process but viewed at different perspectives from Earth. These galaxies have massive black holes at their cores, which are consuming matter from a surrounding accretion disk. Viewed edge-on you see a Seyfert-like phenomenon, but viewed face on you are looking down the axis of a high-speed jet of plasma that changes its brightness rapidly to produce the 'BL Lac' and Blazar phenomena.

One physical system can display many different types of activity, depending on viewing angle.

BLACK HOLES

One of the most peculiar predictions from General Relativity is that highly compressed matter will distort spacetime and cause light to be trapped. These objects, called 'black holes' – a term usually credited to physicist John Wheeler in 1967 – can come in any size from microscopic *quantum black holes* to *stellar-mass black holes*, and even *supermassive black holes* that are millions or even billions of times the mass of a normal star. This is because gravity has no preferred scale of operation, so it is only necessary to compress enough matter to an adequate density for these objects to form at any mass.

Although derived from Einstein's Theory of General Relativity, they were not predicted to exist by Einstein, but instead were gradually discovered in the mathematics of 'point masses' by Karl Schwarzschild in 1916. What he found was that at a specific distance surrounding such compressed masses, the equations would predict the complete trapping of light. This radius became known as the 'Schwarzschild Radius'.

Between 1958 and 1967, regarded as 'the golden age of black holes', many physicists and mathematicians explored the detailed behaviour of these objects for various configurations, including rotating black holes, charged black holes and other possibilities. Astronomers also investigated how they might form in the universe today as a consequence of the normal evolution of stars.

The Schwarzchild Radius

$$R_s = \frac{2GM}{c^2}$$

For stars with masses up to about five times our Sun's mass, the core of a red giant star is at near-white dwarf densities. After the star evolves into the planetary nebula phase and loses its outer envelope of gas, the cooling white dwarf core remains behind. This is the case observed in some planetary nebulae such as M 57 in Lyra. For stars beyond a mass of ten solar masses, the star becomes a supernova. The remnants they leave behind after core collapse can be stable neutron stars if no more than three solar masses are left behind after the detonation. Stars more massive than about 20 solar masses also become supernovae, but can leave behind remnants that are too massive to form stable neutron stars and so collapse directly into black holes. These stellar-mass black holes can form with masses from a few times our Sun's mass up to about 20 times for the most massive 'hyperstars' known to exist today, such as those found in the Eta Carina nebula.

The most massive stars known, of which there may only be a few hundred in an average-sized galaxy, are typically about 100 times the mass of our Sun. These explode as hypernovae in a complex process that also gives rise to black holes, with masses from 10–50 solar masses, along with an intense pulse of gamma radiation. Astronomers can detect these 'gamma ray bursts' from galaxies located several billion light years from Earth. They occur about once each day on average, and represent the most powerful known outbursts of energy in the visible universe.

The Eta Carina nebula shows a massive hyperstar ejecting gas into dumbbell-shaped clouds.

ACCRETION DISKS

Because gravity is scale-free, there is no upper limit to the mass of a black hole. However, as the mass increases, the environment surrounding the black hole becomes more complex. Black holes can gravitationally affect interstellar gas and stars billions of miles from their location. Because of the conservation of *angular momentum*, the accreting matter first takes up residence in a vast, rotating *accretion disk* that can be thousands of kilometres in diameter, as large as our solar system, or even light years across for the most massive black holes. The disks behave like miniature solar systems for which Kepler's Third Law, which we encountered on pages 19–21, gives a relationship between distance and rotation period, but these disks are not stable. Instead, through friction, matter steadily flows inward from the capture location until it reaches the event horizon of the black hole at its centre and then vanishes.

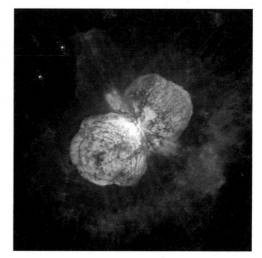

ANGULAR MOMENTUM ▶ *the level of spinning movement in a rotating body. It is 'conserved' (remains the same) unless acted on by another force.*

EVENT HORIZON ▶ *boundary of a black hole, beyond which nothing can escape – not even light.*

ACCRETION DISK ▶ *stellar debris, including gas and dust, which has been pulled into a flattened band of spinning matter around a black hole.*

Black holes are surrounded by accretion disks whose light and radiation make them visible at great distances.

As the matter flows inwards and converts gravitational potential energy into kinetic energy, the accretion disk steadily heats up as you approach the black hole. In the outskirts, the temperature may be only a few hundred K so that the material emits strongly at infrared wavelengths. But near the black hole, temperatures can soar above 100,000 K, making the interior a strong X-ray source. Solar-mass black holes with accretion disks can be detected as X-ray sources. More massive black holes become powerful radio sources as the heated X-ray plasma interacts with the magnetic fields in the accretion disk to generate radio waves. From a poorly understood process, these disks also help *collimate* (focus or align) plasma into beams of energy perpendicular to the disks. These appear as jets of streaming matter, and can also give rise to powerful radio sources at great distances from the host galaxy.

SUPERMASSIVE BLACK HOLES

In 1964, soon after the first quasars were detected, Edwin Salpeter and Yakov Zel'dovich proposed that the ultimate energy for quasars was a 'supermassive' black hole surrounded by an accretion disk. To generate the trillions of solar luminosities of energy inferred for quasars, an entire star like our Sun would have to be consumed each year to liberate the rest-mass energy via Einstein's $E=mc^2$. This proposal was not taken seriously until many separate lines of evidence became available in the 1970s and 1980s. This evidence not only demonstrated that black holes existed but that the energy production in many different *active galactic nuclei* could only be simply explained by the presence of massive black holes millions of times the mass of our Sun. The black hole in the core of Cygnus A (one of the most powerful radio sources in the sky) must have a mass of nearly three trillion Suns to account for the luminosity of this source. A number of quasars such as 3C 175 have prominent jets connecting double radio sources, so the relationship between supermassive black holes in quasars can also be connected with similar objects powering the cores of radio galaxies.

Edwin Salpeter

Edwin Salpeter was an Austrian-American astrophysicist born in Vienna in 1924 and immigrated with his family to Australia, where he completed his bachelor's and master's degrees by 1945. He completed his PhD at the University of Birmingham, England in 1948, and spent the rest of his life at Cornell University. There, he made important contributions to theoretical physics and quantum field theory, as well as astrophysics. He was the first to discover that high-mass stars could convert helium into carbon in the 'Triple-Alpha' process. In 1964 he, and independently Yakov Zel'dovich, proposed that massive blackholes would be surrounded by orbiting disks of matter called accretion disks. From these disks, enormous quantities of radiation would be generated with enough luminosity to account for quasars.

ACTIVE GALACTIC NUCLEI ▶ *the region at the centre of a galaxy with higher luminosity.*

Once the Hubble Space Telescope began its studies of quasars, direct imaging showed that the quasar phenomenon was a nuclear phenomenon among resolvable galaxies that in most cases displayed evidence of collisions in progress. This supported the theoretical idea that supermassive black holes formed as galaxies collided and their nuclear black holes merged to form still more massive black holes. Also, it became clear from a study of the supermassive black hole at the centre of the Milky Way, that perhaps all large galaxies have central black holes that in most cases are dormant, but can flare into Seyfert or quasar-like luminosities when galaxies collide, providing the black holes with fresh fuel.

RADIO ASTRONOMY

For millennia, the only access humans had to the cosmos was through that narrow band of wavelengths called the visible spectrum, extending from short-wave blue and ultraviolet light to the longer wavelengths of red light. In the scope of the entire accessible electromagnetic spectrum, this was like confining your music to one key on a piano. Following James Clerk Maxwell's discovery of electromagnetic waves, there were many attempts between 1896 and 1900 to detect radio waves from the sun but to no avail. During the 1930s, radio engineer Karl Jansky used a simple directional, radio antenna to track down the sources of short-wave radio interference. He eventually found a strong signal that was not the sun, but the centre of the Milky Way galaxy itself. In 1937, and intrigued by Jansky's findings, the amateur radio operator Grote Reber built a 9-m (30-ft) parabolic dish in his backyard and not only detected 'Sagittarius A', but mapped the sky for other radio emissions. He tried frequencies at 3,000 MHz, and 900 MHz before making his first detection at 160 MHz. For nearly a decade afterwards, he published many sky maps as part of his ongoing surveys and was the only practising 'radio astronomer' during this time.

Direct measurement of the orbits of dozens of stars near Sgr-A* indicates that this location is occupied by a massive object with over four million times the Sun's mass. Its optical invisibility, but enormous gravitational influence extending many light years, indicates that this is a classic supermassive black hole. It is currently not able to consume much matter because it seems the interstellar medium is too dilute to supply enough mass over time, and so it is in a dormant state. However, it may have been in an active Seyfert state as recently as two million years ago. The NASA, Fermi Gamma Ray Observatory in 2010 discovered two intense lobes of radiation located just above and below the nuclear core of the Milky Way, centred on Sgr-A*, with estimated ages of about two million years. In addition, clouds of fast-moving gas flowing away from the centre have also been discovered. The Event Horizon Telescope project, an international team of astronomers using radio interferometry techniques, may provide the first glimpse of the event horizon of this black hole. Their project is expected to give radio images of matter falling into the black hole, and the gravitational distortions of the radio light from the intense gravity near the event horizon of Sgr-A*.

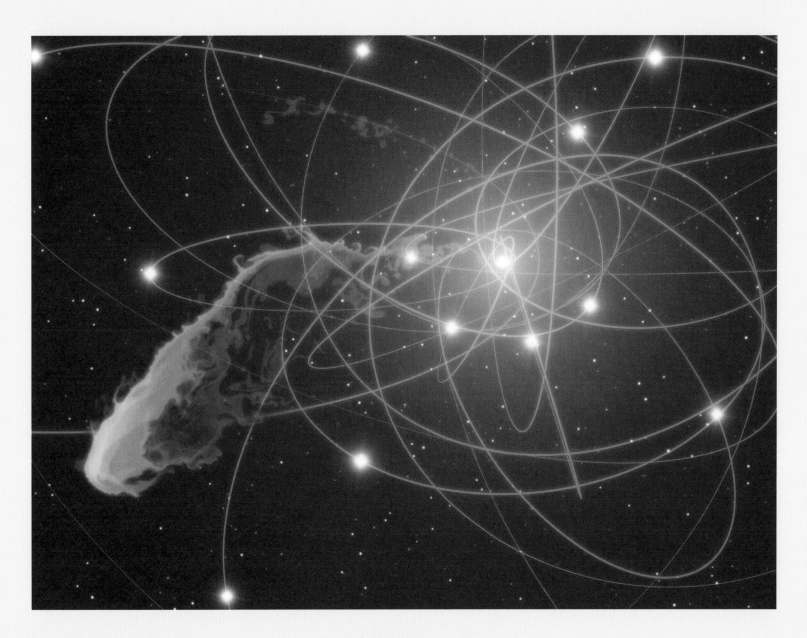

An artist's rendering of the orbits of several dozen stars near Sgr A*, which have been tracked over a decade to determine the mass of the black hole. The red cloud is a computer model of a recent encounter between the black hole and a disrupted interstellar cloud.

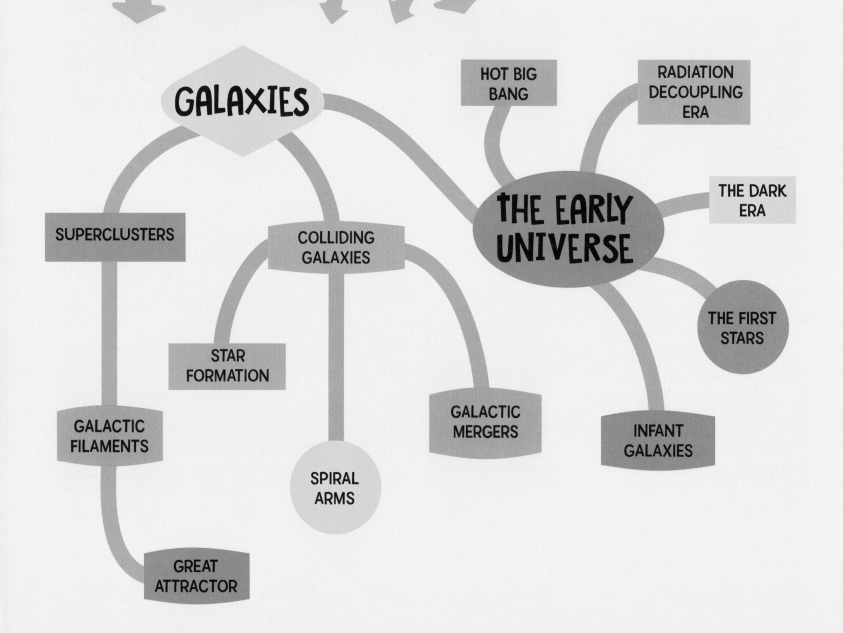

GALAXIES

HOT BIG BANG

RADIATION DECOUPLING ERA

THE EARLY UNIVERSE

THE DARK ERA

SUPERCLUSTERS

COLLIDING GALAXIES

THE FIRST STARS

STAR FORMATION

GALACTIC MERGERS

INFANT GALAXIES

GALACTIC FILAMENTS

SPIRAL ARMS

GREAT ATTRACTOR

COLLIDING GALAXIES

Even in the vastness of extragalactic space, galaxies in motion occasionally encounter each other. This phenomenon is especially common within clusters of galaxies where dozens or even thousands of galaxies may be crowded into volumes of space only a few million light years across. Our own Milky Way has been involved in several collision events with nearby dwarf galaxies. These galaxies have been absorbed into the stellar bulk of our own galactic system in a process astronomers call 'galactic cannibalism'. In approximately two billion years, the nearby Andromeda Galaxy and the Milky Way are destined to collide, which will dramatically transform the shape and contents of the Milky Way.

The colliding galaxies NGC 2207 and IC 2163.

The process of collision and merger has been studied in great detail by looking at actual examples of collisions and cannibalism in distant regions of the sky, and through supercomputer simulations. The gravitational encounters among systems of comparable size are dramatic, and lead to a vast catalogue of shapes, many of which have been observed among actual galactic systems. Glancing encounters cause galaxies to develop spiral arms, while mergers cause the combined stellar system to inflate into elliptical galaxies. When galaxies contain a rich medium of interstellar clouds, the cloud-upon-cloud collisions that result trigger intense bursts of star-forming activity. When the collision speeds are too great, the interstellar medium can be swept out of the galaxies entirely, leaving behind only the original stars, and freezing for eternity the production of newer generations of stars.

Colliding Galaxies.

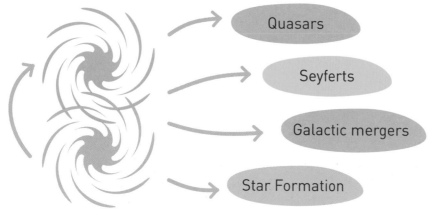

- Quasars
- Seyferts
- Galactic mergers
- Star Formation

Glancing encounters cause galaxies to develop spiral arms, while mergers cause the combined stellar system to inflate into elliptical galaxies. When galaxies contain a rich medium of interstellar clouds, the cloud-upon-cloud collisions that result trigger intense bursts of star forming activity. When the collision speeds are too great, the interstellar medium can be swept out of the galaxies entirely, leaving behind only the original stars, and freezing for eternity the production of newer generations of stars.

Among the first observational programs undertaken by the Hubble Space Telescope was to take images of a number of nearby quasars. These objects for decades had remained completely unresolved star-like objects, and so the premier astronomical question was what they looked like. In 1995, the quasar imaging program led by Princeton University astronomer John Bahcall gave a dramatic answer. Nearly every quasar studied at high resolution showed evidence for a colliding system of two or more galaxies within a very small volume of space. The quasar phenomenon was clearly identified with the core of one of the galaxies, consistent with a collision and merger event taking place. The quasar phenomenon was entirely consistent with a supermassive blackhole being suddenly fed

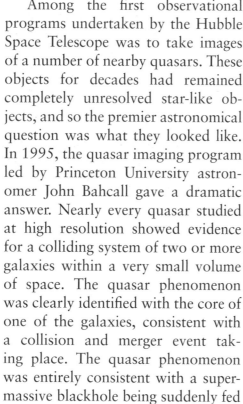

Stefans Quintet is a dense group of spiral galaxies destined to merge into one large elliptical galaxy.

LARGE-SCALE STRUCTURE

Great Attractor

Virgo Supercluster
100 clusters – 110 million light years

Virgo Cluster
1300 galaxies – 54 million light years

Local Group
54 Galaxies – 10 million light years

Milky Way
1 Galaxy – 100,000 Light Years

Galaxies and clusters make up a filamentary universe surrounding large dark voids absent of matter.

by the influx of new gas and dust from the disturbed orbits of one or both of the galaxies.

Some of the most interesting objects in the universe, especially quasars and other active galaxies such as Seyferts (see page 147), appear to be the product of recent or ongoing collisions in which the massive black holes at the centres of these galaxies are ingesting matter. Over time, these supermassive black holes sweep out all the matter whose orbits intersect the black hole and the feeding process stops. However, during a subsequent collision event, new matter can be injected into these orbits and again the supermassive black hole becomes active at a level determined by the inflow rate of the matter.

Galaxy collision and merger is also an important mechanism for growing small galaxy fragments into larger galactic systems. It is thought that this was an important mechanism when the universe was very young and all that existed were small star-forming clouds

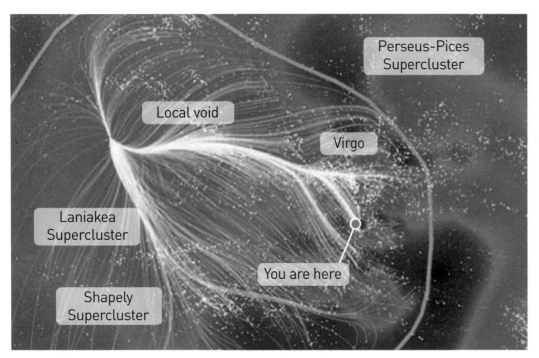

Perseus-Pices
Supercluster

Local void

Virgo

Laniakea
Supercluster

You are here

Shapely
Supercluster

The proposed figure of the local supercluster velocity field .

similar in size to the Magellanic Clouds near our own Milky Way. In a denser and more crowded universe, a larger fraction of these fragments were on collision trajectories, so mergers and cannibalism were far more common during the first few billion years after the Big Bang. Consequently, active galaxies and quasars were much more common in the infant universe, and today there are few nearby examples of these collision-triggered phenomena.

In addition to clusters of galaxies, the universe contains other much larger structures, with individual galaxies as their building blocks. Our Milky Way is a member of the Local Group of galaxies consisting of about 54 individual galaxies, including the Andromeda Galaxy and numerous dwarf galaxies. This group is a member of the Laniakea Supercluster (also called the Virgo Supercluster), which contains more than 100 other clusters of galaxies and spans a roughly spherical volume 110 million light years across. The Laniakea Supercluster is believed to be one of about ten million other superclusters in the visible universe. Along with many other superclusters, it is moving in the direction of the so-called Great Attractor, coincident with the Vela Supercluster. The entire system of galaxies and superclusters is called the Shapley Supercluster and is about 500 million light years across and contains an estimated 100,000 galaxies.

Membership in these vast collections of matter is based not only on the measured recession velocities of hundreds of thousands of individual galaxies, but on supercomputer models of how these galaxies should flow via their velocity fields as a result of their mutual gravitational interactions and the contribution of dark matter. For now, this research remains data-starved. Of the estimated 250 million galaxies within two billion light years of the Milky Way, only a few million have had their redshifts measured. It is estimated that this volume contains as many as one thousand superclusters. In the larger visible universe, perhaps two trillion galaxies remain to be catalogued. Nevertheless, our limited vantage point still lets us explore the basic features of this universe.

A dominant feature of the large-scale galaxy population is its seeming confinement into vast filaments surrounding large galaxy-free voids like the film on the surface of soap bubbles. The specific shape and characteristics of this filamentary structure have been modelled using supercomputer simulations including dark matter, and the observed characteristics match the models if dark matter is the dominant source of gravity.

GRAVITATIONAL WELL ▶ *the pull of gravity exerted by a large body in space.*

During the Big Bang, dark matter formed gravitational 'wells' into which normal matter settled. These wells formed the basis for the large-scale structures we see. As the ordinary matter continued to cool and evolve, it formed individual galaxy fragments that merged to create the populations of galaxies which now constitute the clusters, superclusters and filamentary structure seen today.

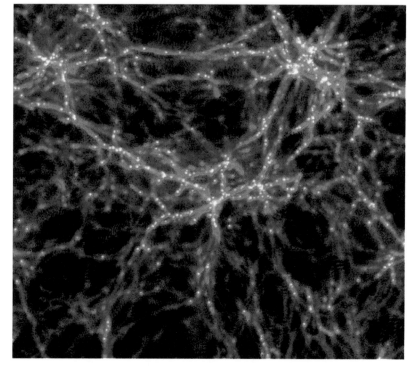

A supercomputer calculation of the evolution of filaments in the universe after the Big Bang.

COSMIC BACKGROUND RADIATION (CBR)

In the late 1940s, American physicists George Gamow and Herman Alpher, both at George Washington University, considered what would happen if the Big Bang were run in reverse, and soon recognized that the infant universe was so hot and dense that nuclear reactions would take place. Gamow assigned the task of applying his mathematical treatment of element formation in the Big Bang to his graduate student Herman Alpher to do the element abundance calculations. These results were finally published in 1948.

This 'hot Big Bang' model was further developed and enlarged during the 1960s and 1970s to include a bewildering number of phenomena that completed their tenure in cosmic history before the cosmos was ten minutes old. If you want to understand what the 'birth' of the cosmos looked like, you are obliged to explore environments known only to high-energy physicists.

Cosmic temperature relation
Because the cosmic scale factor a(t) is related to redshift according to

$$a(t) = \frac{1}{1+z}$$

and because temperature is related to the scale-factor by T a = constant, with the current temperature T = 2.7 K, we get

$$T = 2.7(1+z)$$

What this means is that the CMB was last in contact with ionized matter at z = 1100 T = 2.7(1101) = 3,000K. From the first equation, this happened about 360,000 years after the Big Bang and is called the *Radiation Decoupling Era*.
   Another way to describe these conditions is in terms of the energy of the light in the CMB. This is given by the formula

$$E = \frac{860 \text{ keV}}{\sqrt{t}}$$

where E is the energy of the light in thousands of electron volts. At the time of the start of the Radiation Decoupling Era, t = 380,000 years or 10^{13} s, so E = 0.3 electron volts. This is slightly less energy than ordinary starlight, so the universe at that time would have the brilliance of a reddish star with a surface temperature of 3000 K, and there would have been insufficient energy in these CBR photons to keep the universe's matter ionized.

   In the discussions to follow, we will use the abbreviation CBR for cosmic background radiation, which is the population of photons produced by the Big Bang within which matter is embedded as the universe expands. The equations of Big Bang cosmology allow us to probe this era by defining rather explicitly how the temperature, density and scale factor of space change with time. Because radiation pressure from the CBR is far stronger than the pressure exerted by matter, this period is called the *Radiation-dominated Era*. This era ended at about the time that the CBR last had contact with matter at about 380,000 years after the Big Bang (z = 1100) in what is called the *Radiation Decoupling Era*. At that time, matter pressure became more important to controlling the expansion of the universe than the pressure of the CBR radiation. Since then, we live in the Matter-Dominated Era.

CBR ▶ *cosmic background radiation: the light energy produced by the Big Bang. The CBR has cooled over time and is now only detectable at microwave wavelengths, hence cosmic microwave background (CMB) radiation.*

The CMB imaged by the WMAP program, showing complex structure.

WHAT WAS THE DENSITY AT THAT TIME?

The current density of baryonic matter in the universe is about 1 proton for every four cubic metres of space or 4×10^{-31} g/cc. At a redshift of 1100, the scale factor for space is 1100 times smaller than today. That means the density at this earlier time is $(1100)^3$ or 1.3 billion times more compressed, giving us a matter density at that time of about 5.3×10^{-22} g/cc. That works out at 1.3 billion protons in every four cubic metres of space.

As the universe continued to expand, the temperature of the CBR continued to decline. Before Z=1100, this radiation carried enough energy that it kept the plasma ionized at that time. This consisted of electrons, protons, and the nuclei of deuterium, tritium and helium. But after this redshift, the CBR radiation carried insufficient energy and so the ionized nuclei attracted the appropriate number of electrons to become neutral atoms.

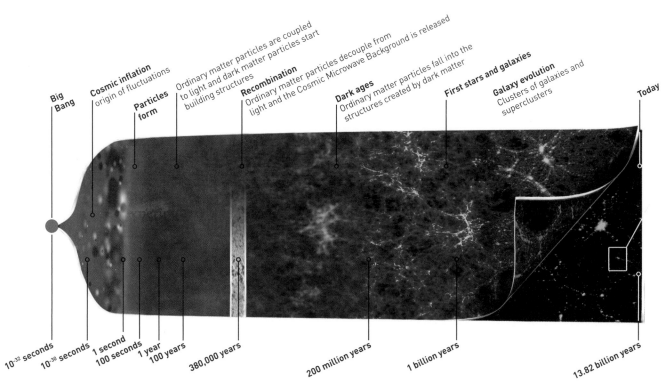

Big Bang

Cosmic inflation
origin of fluctuations

Particles form

Ordinary matter particles are coupled to light and dark matter particles start building structures

Recombination
Ordinary matter particles decouple from light and the Cosmic Microwave Background is released

Dark ages
Ordinary matter particles fall into the structures created by dark matter

First stars and galaxies

Galaxy evolution
Clusters of galaxies and superclusters

Today

10^{-32} seconds 10^{-30} seconds 1 second 100 seconds 1 year 100 years 380,000 years 200 million years 1 billion years 13.82 billion years

A graphical timeline of the early history of the universe.

THE DARK ERA

As the universe continued to expand and cool, the temperature of the CBR radiation continued its decline until by a redshift of $z = 100$ at a temperature of about 270 K, the only traces of this cosmic background radiation had disappeared as visible light and became the much cooler infrared radiation. It would eventually continue to cool to 2.7 K and be detected at the current time as the cosmic microwave background (CMB) discussed earlier. Astronomers call this the start of the Dark Era in cosmic history. It began about 40 million years after the Big Bang and continued until the first stars formed in the universe, believed to have started about 200 million years after the Big Bang.

During the Dark Era, the distribution of the atomic hydrogen and helium gas was gravitationally decided by the pattern of gravitational wells defined by dark matter.

For every gram of normal matter there was about 5 g of dark matter. So, the large-scale irregularities in normal matter, leading to clusters and superclusters of galaxies, were dictated by the gravitational irregularities in the dark matter.

As the hydrogen gas continued to cool, it fell below the escape velocities of these dark matter gravity wells and became clumped. These clumps contained upwards of hundreds of times more mass than the Milky Way galaxy and provided enough matter to subsequently clump to form galaxy fragments within these dark matter wells.

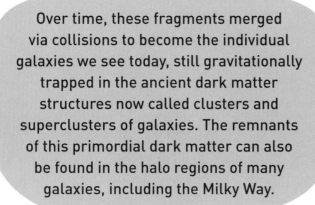

Over time, these fragments merged via collisions to become the individual galaxies we see today, still gravitationally trapped in the ancient dark matter structures now called clusters and superclusters of galaxies. The remnants of this primordial dark matter can also be found in the halo regions of many galaxies, including the Milky Way.

THE FIRST STARS

The processed ejecta from these supernovae were enriched by nuclear fusion into many of the elements in the periodic table, such as the easily detectable elements iron, carbon and oxygen. These heavier elements were now appearing for the first time in cosmic history, and were incorporated into the Population II stars. The Population III stars were hot and intensely luminous at ultraviolet wavelengths, which ionized all the hydrogen in their vicinity out to several hundred light years. This became

An artist's rendering of a Population III star being born and ionizing its environment.

the first time that ultraviolet light had appeared in the universe since the Big Bang. As it flooded the universe, the universe entered a second phase, which astronomers call the Epoch of Reionization. As the Reionization period proceeded, the stray intergalactic hydrogen and helium gas became ionized, producing a dilute hot intergalactic medium that persists to the present time. Also, the dark pre-stellar clouds that survived being evaporated remained to be detected in the spectra of distant quasars and are called 'Lyman-alpha clouds'.

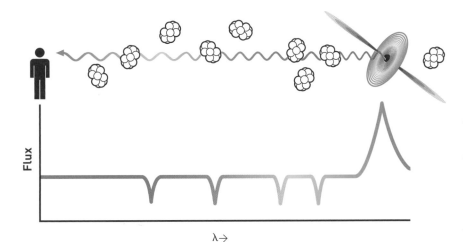

Each cloud along the line of sight to a distant quasar produces a specific absorption feature in the quasar's spectrum at a specific redshift (distance). These absorption features can be so numerous that they are called the Lyman-Alpha Forest in spectroscopy.

Early Stars of the Milky Way

Type	Population I	Population II	Population III
Age	Young stars	Older stars	First generation of stars
Composition	Rich in heavy elements	Poor in heavy elements	Only hydrogen, helium and traces of deuterium
Mass	>= 1 Solar Mass	> 1 solar mass	100–500 solar masses
Location	Spiral arms	Globular clusters	Throughout the universe
Example	Our Sun	SM0313	Hypothetical

Ancient star

We don't have to look into distant space to see the oldest stars that formed. In 2017, astronomers discovered one of the first stars to form in the Milky Way. The star called SM0313 is located 6,000 light years away in the halo of the Milky Way and likely formed just 100–200 million years after the Big Bang, some 13.6 billion years ago. It has significant traces of carbon, so it must have formed after the Pop III stars in its vicinity had exploded and processed the pristine primordial hydrogen and helium into detectable carbon and other elements..

INFANT GALAXIES

At each age, the temperature of a gas and its gravitational dynamics sets a scale at which matter becomes unable to support itself and collapses into discrete clouds of matter. The cooling gas collects preferentially in the gravitational wells formed by dark matter. At about 100 million years after the Big Bang, these clouds had masses from 100,000 to several million solar masses. As they collided and merged, they built up larger collections of matter called *protogalaxies*. Eventually star formation began in these protogalaxy cores, making them visible for the first time. Remnants of this protogalaxy population may exist today in the form of irregular galaxies such as the Magellanic Clouds.

During the Dark Era, the distribution of the atomic hydrogen and helium gas was gravitationally decided by the pattern of gravitational wells defined by dark matter.

Because collisions were frequent, star formation was an intense process for some of these protogalaxies, which formed the earliest populations of stars very quickly. Today, we see these very old stellar systems as elliptical galaxies and dwarf elliptical galaxies. They typically contain only Population II stars, poor in heavy elements. For other protogalaxies, the star formation process was slower and prolonged, leading to spiral-type populations today. These galaxies contain a mix of Population I and II stars along with interstellar clouds in which star formation can still occur. Thanks to the effect of gravitational lensing (see pages 66–7) and the combined efforts of the Hubble Space Telescope and other instruments, astronomers have detected the dim images of many protogalaxies and have started the process of investigating many of their properties.

In 2016, astronomers using the Hubble Space Telescope spectroscopically confirmed the farthest galaxy known by that time, called GN-z11. It is located at a redshift of z=11.1. The light that forms its dusky image began its journey a mere 400 million years after the Big Bang, placing it near the end of the Dark Age, and at the start of the age of galaxy formation in the universe. It is about $^1/_{25th}$ the size of our Milky Way and contains about 1 per cent of the Milky Way's mass, but is forming stars at a pace 20 times faster than the Milky Way. The very rapid pace of star formation suggests that it is destined to become part of an elliptical galaxy.

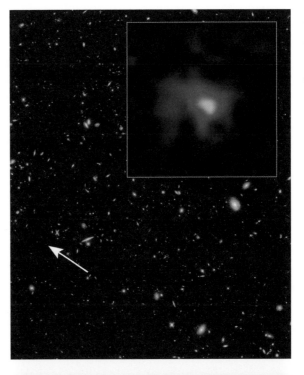

An image of a very young galaxy, EGS-zs8-1, which formed just 670 million years after the Big Bang.

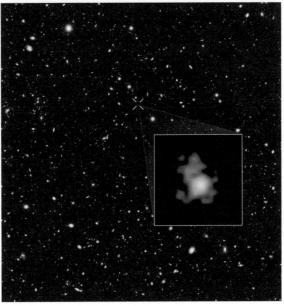

The farthest known galaxy, GN-z11.

PLOTTING THE COSMIC MICROWAVE BACKGROUND

Starting in 1989, three spacecraft of increasing sophistication have been launched to map the cosmic background radiation at microwave wavelengths (CMB) where its emission is the brightest. These craft were called COBE, WMAP and Planck. They also determined an ultra-precise temperature for this radiation, and mapped out its variation across the entire sky at a resolution of a few arcminutes.

The first feature to be recognized comes from the Doppler shift of our combined solar system and Milky Way motion through space, and it is easily seen in the all-sky maps. In the direction of this motion towards the constellation Aquarius, the entire sky hemisphere appears slightly hotter than in the trailing hemisphere. The difference in temperature of 0.0033 K is directly related to the Doppler shift of 600 km/s in the direction of the constellation Aquarius: the nominal location of the Great Attractor.

If this 'dipole' Doppler effect is removed and the average temperature of the CMB – 2.726 K – is subtracted, you are left with a difference map that emphasizes the variations in the CMB across the sky. This is called an *anisotropy map*, and it reveals important information about the way that matter was distributed in the early history of the universe.

When the cosmic background radiation travels through the gravitational well of a collection of matter, it loses some of its energy in direct proportion to the depth of the gravitational well. The temperature variations detected in the CMB map are at a level of $1/100,000$ of the 2.726 K, which also confirms that the Big Bang was a highly uniform event as is assumed by Big Bang cosmology. The pattern of these variations (called *anisotropies*) represents a fingerprint left in the CMB by the distribution of dark matter and ordinary matter.

If we compared the temperature of the CMB at two points located a fixed distance apart in the sky (measured in degrees) and did this for all possible such pairs of points across the sky, we can construct what is called the anisotropy power spectrum of the CMB.

The resulting power spectrum is a snapshot of the pressure changes (sound waves) taking place at a variety of wavelengths at the time the universe became transparent to the cosmic background radiation, which happened about 380,000 years after the Big Bang. The intensity and wavelengths of these modes revealed in the anisotropy power spectrum is directly connected to the physics of the early universe. Detailed study of the first peak, its shape, wavelength and intensity, reveals information about the geometry of spacetime, namely that the universe is very close to being flat. The ratio of odd to even peaks indicates the universe's total baryon density, while the location of the third peak yields clues about the density of dark matter.

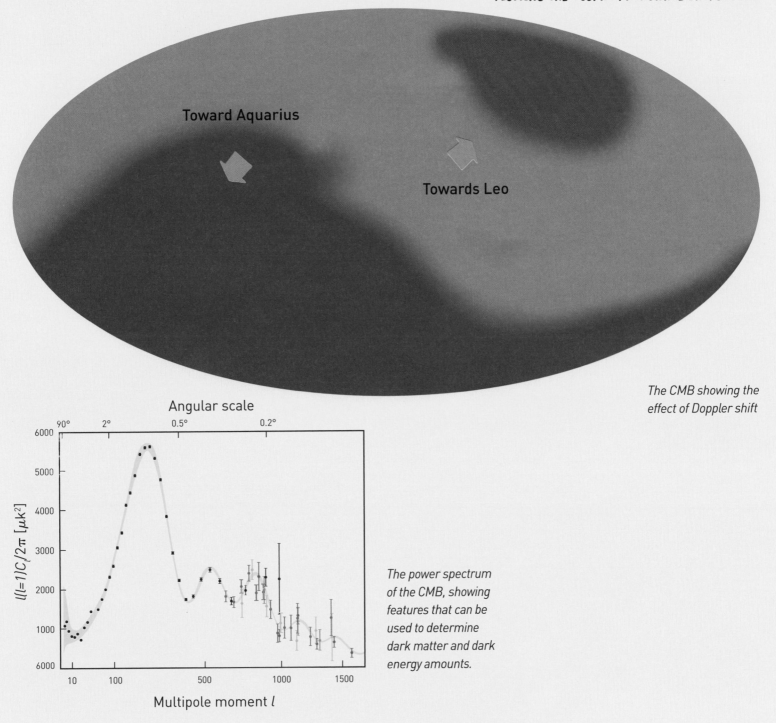

Toward Aquarius

Towards Leo

The CMB showing the effect of Doppler shift

The power spectrum of the CMB, showing features that can be used to determine dark matter and dark energy amounts.

Angular scale

90°   2°   0.5°   0.2°

$l(l=1)C_l/2\pi$ $[\mu K^2]$

Multipole moment l

Chapter Ten
THE ORIGIN OF THE PRIMORDIAL ELEMENTS

Primordial Element Abundances – The Nucleosynthesis Era – The Lepton Era– The Quark Era – The Electroweak Era – Matter-Antimatter Asymmetry

NEUTRONS AND PROTONS

PRIMORDIAL

QUARK ERA

ELECTROWEAK ERA

CREATION OF THE ELEMENTS

THE BIG BANG

HEAVY

LEPTON ERA

NUCLEOSYNTHESIS

HYDROGEN AND HELIUM

GUT ERA

CMB RADIATION

BARYON TO ENTROPY RATIO

MATTER-ANTIMATTER ASYMMETRY

SAKHAROV CONDITIONS

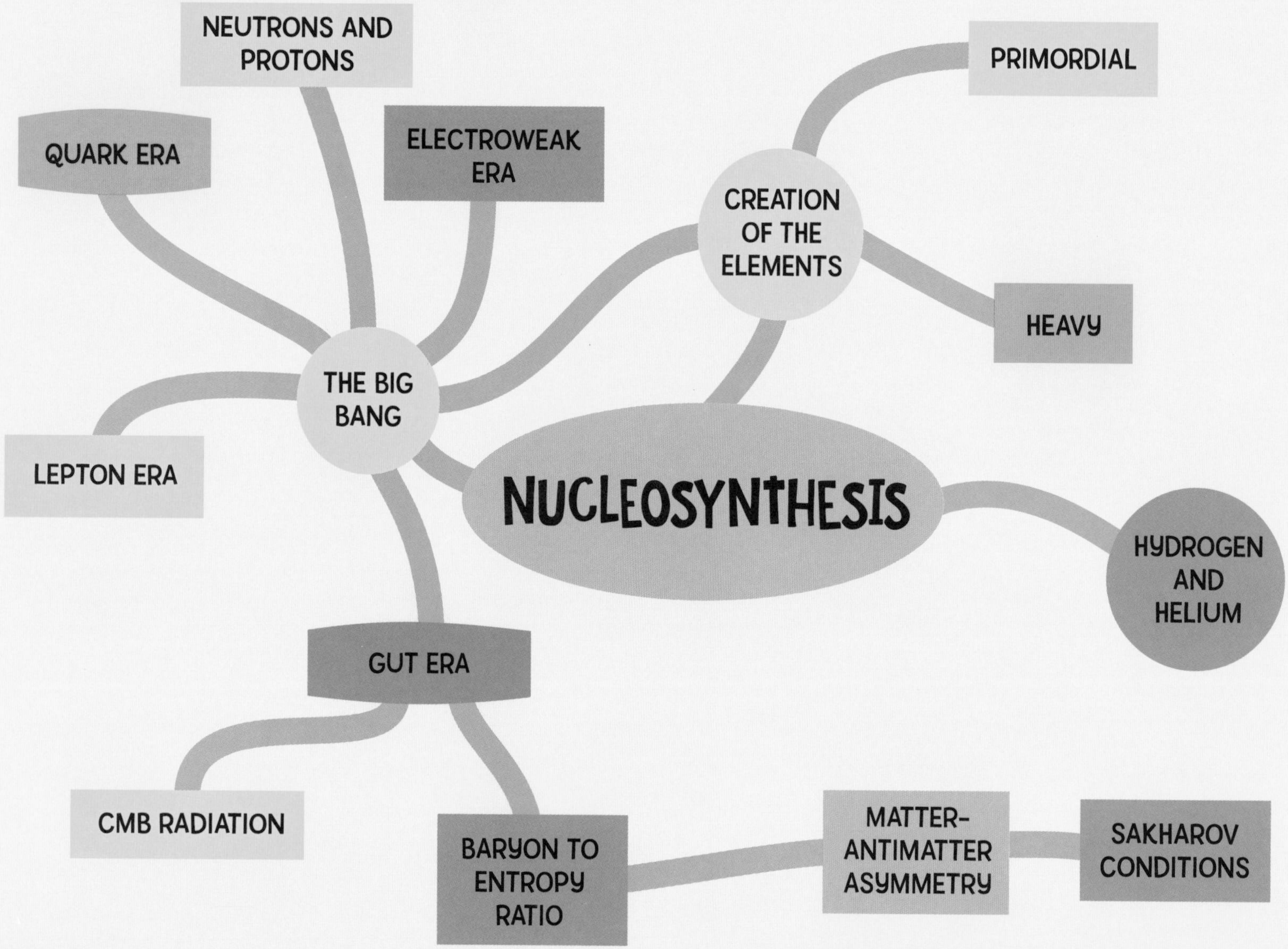

PRIMORDIAL ELEMENT ABUNDANCES

Since the 1960s, astronomers have measured the abundances of the elements in the periodic table and developed the idea that there are two categories of elements in the universe. In the first category, called the primordial elements, we have hydrogen, deuterium, tritium, helium, lithium and beryllium. In the second category, called the heavy elements, we have all the others.

PRIMORDIAL ELEMENTS ▶ *hydrogen, deuterium, tritium, helium, lithium and beryllium.*

HEAVY ELEMENTS ▶ *all the elements apart from the primordial elements.*

As discussed previously, heavy elements are believed to be formed in the cores of evolving stars, and ejected into space by supernova explosions. There they mix with the gas and dust of the interstellar medium and eventually form the next generations of stars. Over time, successive generations of supernovae cause a steady enrichment in the abundance of heavy elements. The earliest stars that formed would have the lowest abundance of heavy elements, while the most recently formed stars are more likely to have higher abundances of heavy elements. The abundances can vary from galaxy to galaxy, and from one region in a galaxy to the next, depending on the level of star formation taking place.

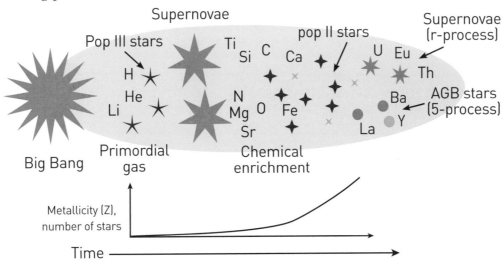

The various stages in the formation of the elements.

Primordial elements, on the other hand, were formed soon after the Big Bang. Their abundances in the early universe can be discerned by looking at the oldest and most pristine stars in a galaxy, which are found in globular star clusters, in the halo population of stars surrounding the Milky Way, and in most elliptical galaxies. About 75 per cent of all primordial elements are hydrogen (1 proton), 25 per cent is helium (2 protons and 2 neutrons) and less than 0.01 per cent deuterium (1 proton, 1 neutron), lithium (2 protons 1 neutron) and beryllium (3 protons, 2 neutrons). A proper cosmological model has to account for the origin and abundances of these primordial elements.

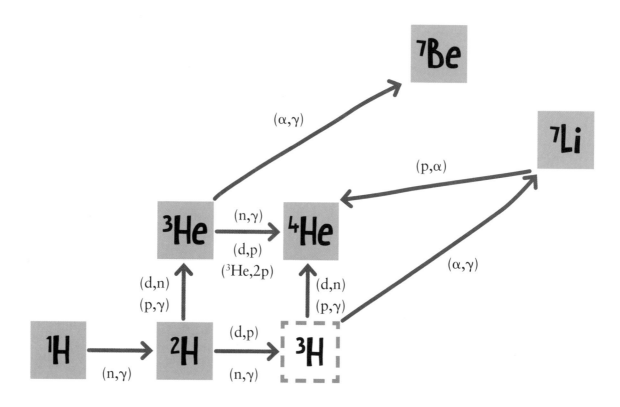

THE NUCLEOSYNTHESIS ERA (3 MINUTES TO 20 MINUTES)

Between three minutes and twenty minutes after the Big Bang, the universe was dense enough, and temperatures were high enough, for the primordial atomic nuclei of deuterium, lithium, tritium and helium to form. At the start of this era some three minutes after the Big Bang, temperatures were above 700 million degrees. A plasma of protons, neutrons and electrons was kept hot by incessant collisions with the individual photons of the CBR. The density of matter everywhere in space was about 7×10^{-6} g/cc. This is about $^1/_{100}$ the density of the air we breathe, but every location within the universe was at this temperature and density. This temperature was at the threshold where the CBR photons had just enough energy to collide and fragment any protons and neutrons that tried to form low-mass stable nuclei.

After about three minutes, protons and neutrons were just able to fuse together to form deuterium, tritium, lithium and helium nuclei because the CBR photons carried far less energy to split apart (fission) the newly-formed nuclei. As the universe continued to expand and cool over the next 15 minutes, the collision temperatures and density of matter became too low for nuclei much heavier than helium to form, and so the element production process stopped near the element beryllium, allowing the abundances of these elements to be frozen in. At this point we had a plasma consisting of these primordial element nuclei, electrons, free neutrons, and CBR photons.

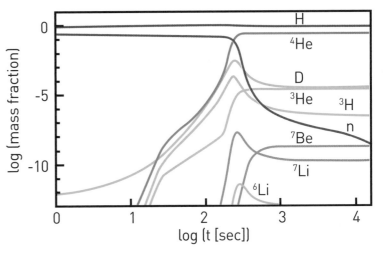

Protons are stable elementary particles but neutrons are not. Neutrons have a half-life of ten minutes, so that ten minutes after the Big Bang, with no way to produce more neutrons, those left over from building the primordial elements soon vanished, leaving behind only the primordial element nuclei: hydrogen, deuterium, lithium, tritium and helium, along with electrons and the photons in the CBR.

The changing primordial abundances as the universe expanded and cooled.

If we wish to end the story of the origin of familiar matter in the Big Bang, we may stop here. We have now solved another cosmological mystery using the Big Bang model: why is the ratio of hydrogen to helium found to be 3 for 1 in nearly every old object we can study across the universe? With reference to Population III stars, we also have a pretty good idea where the elements heavier than helium came from. But mathematically, Big Bang theory predicts how to compute the temperature and density of the cosmos, specifically the CBR, for even earlier events in cosmic history. So we are obliged by the mathematics to inquire about what happened less than three minutes after the Big Bang.

Hydrogen-to-Helium ratio

THE LEPTON ERA (10^{-8} SECONDS TO 1 SECOND)

So, what did happen within the first three minutes after the Big Bang? To answer this question, we have to go well beyond atoms and atomic nuclei and investigate what happens to matter at temperatures above 10 billion K when the universe was only one second old. As physicist Steven Weinberg pointed out in the 1970s, we can only do this if we first understand the deep

Steven Weinberg

Steven Weinberg is an American theoretical physicist and Nobel laureate in Physics for his contributions with Abdus Salam and Sheldon Glashow to the unification of the weak force and electromagnetic interaction between elementary particles. Weinberg was born in 1933 in New York City and attended the Bronx High School of Science. He received his PhD at Princeton University in 1957. His primary research was the investigation of the weak interaction, which led him to studies of spontaneous symmetry breaking and the co-development of electro-weak theory involving the Higgs boson. His textbook *Gravitation and Cosmology* has been required reading for several generations of astrophysicists. In 1977, his popular book *The First Three Minutes: A Modern View of the Origin of the Universe* became a best-seller, along with his laconic end-quote: 'The effort to understand the universe is one of the very few things which lifts human life a little above the level of farce and gives it some of the grace of tragedy,' which stimulated years of religious commentary and reprobation.

structure to matter and energy in our universe. Weinberg was one of the pioneers in making the investigation of the earliest moments in cosmic history a respectable field of investigation, not only through his technical writings but also through the popularization of the subject through his book *The First Three Minutes*.

Fortunately, the groundwork was provided by physicists who developed the Standard Model, which we discussed in Chapter 4. The term 'Standard Model' as we have seen is the physicists' moniker that describes the 13 fundamental particles which make up the matter (electrons, quarks, neutrinos and the Higgs boson) in the universe, together with the three elementary forces mediated by their 12 additional elementary particles, which are responsible for the interactions between matter particles.

From the Big Bang model, when the universe was only about one second old, the temperature of the CBR was near ten billion K. At this time the universe was just hot enough that photons could convert into electron–positron pairs and these pairs could annihilate into photons so that the creation and destruction of these pairs was in balance. As the universe continued to expand and cool, the temperatures after a few seconds fell below this threshold and so the electron–positron pairs started to decay away and vanish from the universe, from that time forward.

During the Lepton Era, the most massive tau leptons were being produced by the CBR. Their masses are 1.8 GeV, so a pair of tau–antitau particles produced by a single CBR photon would require a photon energy of 3.6 GeV. This energy was available when the universe had an age of 6×10^{-8} s and at a temperature of 40 trillion K.

The time between the equilibrium formation of tauons and the final decay of the electron–positron pairs, spanning the time interval from 10^{-8} seconds to about one second after the Big Bang, is called the Lepton Era. This era started out with the average density of the universe near 100 trillion g/cc, which is comparable to the density of large atomic nuclei.

Another event that occurred at about one second after the Big Bang and at the end of the

Cosmic Temperature

Before the cosmos was 380,000 years old, the CBR completely dominated how the universe expanded through its intense pressure. This radiation took the form of a perfect black body (its spectrum is defined by a single parameter: its temperature). The way that the temperature of the CBR changed in time can be computed from the Big Bang theory and follows the formula

$$T = \frac{10^{10}}{\sqrt{t}}$$

he current time after the Big Bang in seconds, and T is the temperature in K. Because the temperature of a system of particles represents their average energy, we can also write this formula in terms of the average energy of the CBR photons according to

$$E = \frac{0.00086}{\sqrt{t}}$$

where the photon energy (E) is given in billions of electron-Volts (GeV). For example, by the end of the Lepton Era, at one second after the Big Bang, the temperature was 10 billion K and the CBR photons each carried about 0.00086 GeV or 860,000 eV of energy. This energy is too low for electron–positron pairs to be produced, so the production of the lightest known leptons (of matter) ended.

Lepton Era was that the ratio of protons to neutrons became fixed at one neutron for every seven protons. At higher temperatures and earlier times, protons could be transformed into neutrons by absorbing an electron. Neutrons, meanwhile, could be converted into protons by absorbing an antielectron. This production and destruction process was in balance before one second because there were copious electrons, antielectrons, neutrinos and antineutrinos being produced by the CBR. The ratio of protons to neutrons was initially fixed at one to one, but as the universe cooled, this equilibrium broke down and fewer antielectrons were produced by the end of the Lepton Era. The ratio shifted to one neutron for every seven protons during what is called the neutron–proton freeze-out period. These neutrons and protons remained free particles until temperatures fell below 700 million degrees K at the start of the Nucleosynthesis Era.

Proton-to-neutron ratio

Neutron-proton freeze-out period

THE QUARK ERA (10⁻¹⁰ TO 10⁻⁶ SECONDS)

As we have seen in Chapter 4, the Standard Model describing how matter and forces interact has been tested at the Large Hadron Collider and found to be accurate up to energies of 13 TeV. This means we can push back our knowledge of the early Big Bang far beyond the one-second limit.

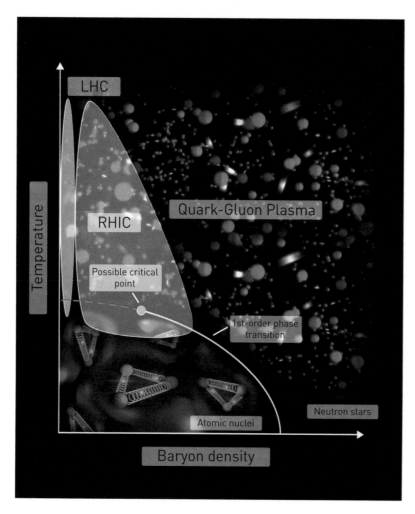

The quark-gluon plasma as temperature and cosmic density change.

We know that protons and neutrons are composed of quarks according to the Standard Model. An interesting feature of the gluon strong interaction between quarks is that the interaction weakens as you compress the protons and neutrons into smaller volumes. The gluons and quarks begin to behave like a gas of weakly interacting particles. By the time the cosmos had achieved densities above 10^{14} g/cc, matter became so compressed that despite the temperature of the CBR, protons and neutrons began to dissolve into a quark-gluon plasma. This happened at an energy above 1 GeV per particle or a temperature of 10 trillion K. This would have been the situation about 10^{-6} seconds after the Big Bang, near the start of the Lepton Era. Quark-gluon plasma are being explored in considerable detail at the Brookhaven Relativistic Heavy Ion Collider, which has helped fill in many details about the equivalent era in cosmic history.

QUARK-GLUON PLASMA ▶ *the compressed state of matter after quarks and gluons which could exist in the extreme temperatures of the Big Bang.*

THE ELECTROWEAK ERA (10^{-12} TO 10^{-36} SECONDS)

The next major event in early cosmic history occurs when the electromagnetic and weak forces begin to look dissimilar. This heralds the end of what is called the '*Electroweak Era*', during which time these forces behave in a nearly identical manner and are numerically indistinguishable in terms of their strengths. From the Standard Model, this transition occurs when the Higgs boson gains mass and its interactions with the other matter fermions and bosons causes them to gain mass to varying degrees. Before this transition, quarks, gluons, leptons, neutrinos, photons and the W and Z bosons were all massless. After this transition, the W and Z bosons gained a lot of mass, while the gluons and photons remained massless.

The details of when this transition happened are still being developed, but given that the observed mass of the Higgs boson is 126 GeV, calculations seem to indicate that the change in the shape of the Higgs potential from the symmetric electroweak vacuum to the current vacuum state occurs between 100 and 300 GeV. Also, the pair production of the W and Z bosons stops once the CBR energy drops below about 160 GeV, so the changeover to the Electroweak Era occurs near a cosmic time of 3×10^{-11} s when the CBR temperature was near 2×10^{15} K. *The end of the electroweak era.*

As the universe continues to expand and age, by the end of the Quark Era, the electromagnetic and weak interactions looked progressively more different as the symmetry between them was broken in a rapidly cooling universe. Before this time, the symmetry became more and more exact as the Higgs boson rapidly lost its mass.

Thanks to the development of the Large Hadron Collider (LHC), the conditions during most of this electroweak era can be simulated and compared to Standard Model predictions. The maximum LHC energy of 13 TeV corresponds to a temperature of 1.5×10^{17} K, which was reached at a time of 4×10^{-15} seconds after the Big Bang. That means our Standard Model, which has been tested and found completely accurate up to this energy, has now opened up the entire history of the Big Bang starting 0.000000000000004 seconds after the Big Bang, deep within the Electroweak Era, and moving forward to the Lepton Era. *Large Hadron Collider*

What did this universe look like?

First, the density of the universe was much greater than the near-vacuum of today of 10^{-31} g/cc to $10^{-31} (1.5\times10^{17}/3)^3 = 10^{19}$ g/cm^3. This is over 100,000 times more dense than the nucleus of an average atom. At these densities, the entire mass of the Milky Way galaxy could be encompassed by a ball of matter about the diameter of our own planet Earth. The Andromeda Galaxy, currently located 2.6 million light years away, was then a similar-sized ball of matter about 500 km (310 miles) away. In other words, the entire system of galaxies we see around us today, out to ten times as far as the Andromeda Galaxy were all smashed together into matter quite close to us at this time. You could travel at a comfortable walking speed from the matter in one galaxy to the other in a few months. *Density of the universe*

So any temperature variation could only be smoothed out if its scale was smaller than the current horizon size at a given time after the Big Bang. This problem leads, today, to the expectation that objects farther apart in the sky than about one degree cannot have been in thermal communication yet because there has not been enough time for the light to be exchanged between these points to smooth out their temperatures. The fact that the COBE and WMAP CMBR data show a nearly smooth temperature across the entire sky means that there has to have been some process taking place smoothing out temperature differences nearly at the instant of the Big Bang itself. This would have happened even before the Electroweak Era.

Early cosmic history demarcated in terms of the major eras.

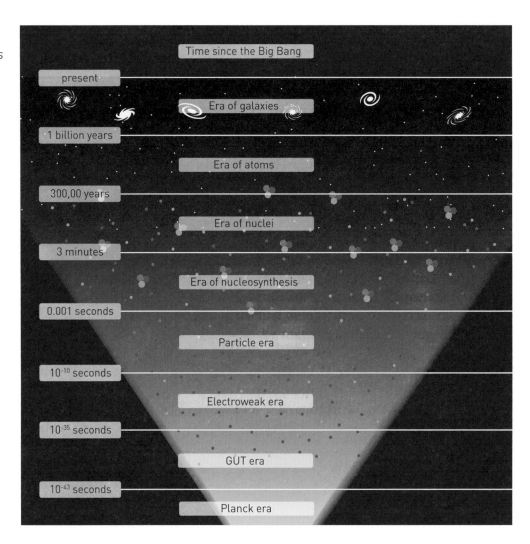

When did the Electroweak Era start?

The next event predicted by extensions of the Standard Model is the unification of all three forces in what is called the Grand Unified Theory (GUT) Era. The GUT Era ended when the strong force became distinct from the electroweak 'unified' force that defines the Electroweak Era. We will cover this important event in the next chapter. As we will see, this transition happened at about 10^{-36} seconds after the Big Bang. So we can estimate that the Electroweak Era began about 10^{-36} seconds to about 10^{-12}

seconds after the Big Bang. Meanwhile, we have at least two unresolved issues to address.

DARK MATTER

We originally said which apart from the dark energy which comprises about 75 per cent of the modern universe, the remaining gravitating material is dominated by dark matter by a factor of nearly five-fold (26 per cent dark versus 5 per cent normal). If this dark matter is in the form of new types of particles, what were these particles doing during the Quark and Electroweak Eras? All of the calculations so far have only included interactions among ordinary Standard Model matter. Nevertheless, from this limited view, our detailed descriptions of particle interaction events since 10^{-15} seconds after the Big Bang have led to accurate predictions for the neutron-proton ratio, the abundance of the primordial elements, and the number of generations of leptons (the abundances are consistent with at most three generations), without any 'corrections' for interactions with dark matter particles. We also know that the dark matter particles, if they are in fact particles, are weakly interacting with the Standard Model particles and are likely to be very massive, exceeding at least several TeV per particle.

Standard Model interactions

This means that by a time of 10^{-15} seconds after the Big Bang, during the Electroweak Era, the production and annihilation of matter-antimatter pairs of dark matter particles by the CBR would have ended, and the residual dark matter particles would thereafter interact only weakly with the Standard Model particles from that time forward. That suggests all of the events beginning with the start of the Quark Era and leading to the Nucleosynthesis Era proceeded as though no dark matter existed. Dark matter particles only served to affect these processes through their gravitational effect upon the rate of expansion of the universe. At some point during the early Electroweak Era, however, dark matter interactions likely became important and may have created an earlier Dark Matter Era before the Electroweak Era began.

Role of dark matter

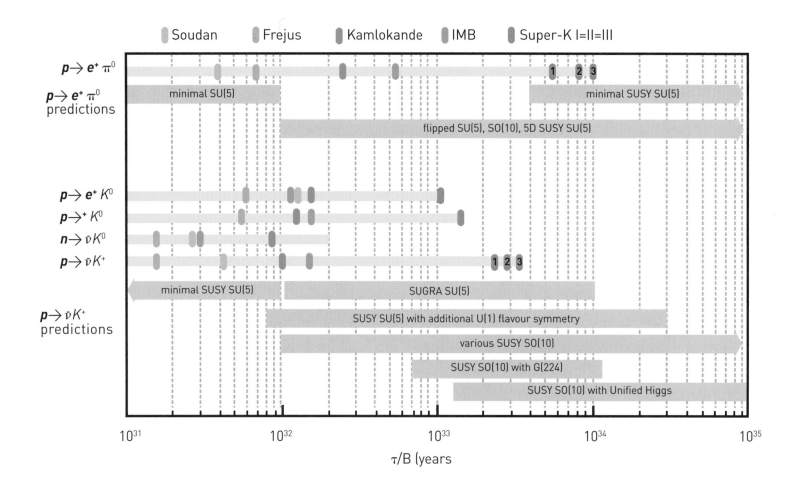

Soudan   Frejus   Kamlokande   IMB   Super-K I=II=III

$p \rightarrow e^+ \pi^0$

$p \rightarrow e^+ \pi^0$ predictions

minimal SU(5)   minimal SUSY SU(5)

flipped SU(5), SO(10), 5D SUSY SU(5)

$p \rightarrow e^+ K^0$

$p \rightarrow^+ K^0$

$n \rightarrow \bar{\nu} K^0$

$p \rightarrow \bar{\nu} K^+$

$p \rightarrow \bar{\nu} K^+$ predictions

minimal SUSY SU(5)   SUGRA SU(5)

SUSY SU(5) with additional U(1) flavour symmetry

various SUSY SO(10)

SUSY SO(10) with G(224)

SUSY SO(10) with Unified Higgs

10^{31} 10^{32} 10^{33} 10^{34} 10^{35}

τ/B (years)

Some limits showing that protons are stable for longer than about 10^{33} years.

MATTER-ANTIMATTER ASYMMETRY

The Standard Model has been tested to energies as high as 13 TeV, so we can take one more step that is still rooted in experimental data. At 4×10^{-15} s, we reach the limits of the Standard Model (c.2018, tested by the LHC). What we now see looking forward to a cooling universe is a dense plenum (a space completely filled with matter) of essentially massless elementary quarks, leptons and bosons together with all of their anti-particles being freely created and destroyed as they interact with the photons from the cosmic background radiation. There are no new particles created in this maelstrom beyond the elementary fermions and bosons of the Standard Model. No other composite particles have any chance of existing as the incessant collisions shatter them apart before they can last longer than the age of the universe at that time. We are, however, left with two very fundamental problems. Where did the CBR come from, and why is there more matter than antimatter in the universe?

The ratio of baryons to photons is called the *cosmological baryon-to-entropy ratio*, and from the work by the WMAP mission, this number is 6.1×10^{-10} baryons/photons. A more careful definition of this ratio that accounts for the changing density of the universe as it expands leads to the *baryon-to-photon density ratio*, whose value is 10^{-8}. This means that for every 100 million antiquarks, there should be 100 million plus 1 quarks.

When you compare the number of photons in the CBR with the number of baryons (protons plus neutrons) in the universe, you discover that there are about one billion photons per baryon. What this means is that instead of there being equal numbers of baryons (matter) and antibaryons (antimatter) in the universe, the baryons outnumber the antibaryons by one part in one billion. This ratio was essentially constant even by the start of the Electroweak Era. Some of these photons were involved in various stages of pair production, but when these events fell out of equilibrium, the total number of photons did not significantly change. In a matter–antimatter symmetric universe, if there had been exactly equal amounts of matter and antimatter, there would be no matter left over to form stars and galaxies, only a dilute and cooling gas of cosmological photons.

The reason for the one part in one billion excess of matter over antimatter is not known, although many mechanisms have been investigated. None of these mechanisms takes place after the end of the Electroweak Era at 10^{-12} seconds, and our current experimental limit set by the LHC shows that nothing unusual is happening to the Standard Model up to 13 TeV corresponding to a time of 10^{-15} seconds after the Big Bang. The processes that led to the one-in-one billion imbalance must have been in operation much earlier than 10^{-15} seconds after the Big Bang. What would these processes have looked like? At least theoretically, it appears that the only way you can create a net baryon excess starting from a system in equilibrium is if the so-called Three Sakharov Conditions, proposed by Soviet physicist Andrei Sakharov in 1967, are satisfied.

Baryon-to-antibaryon ratio

Before

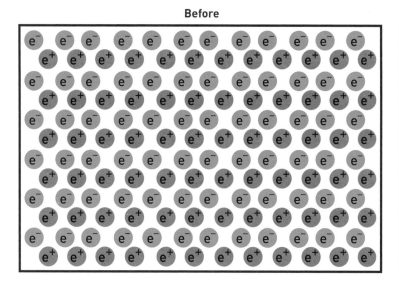

After

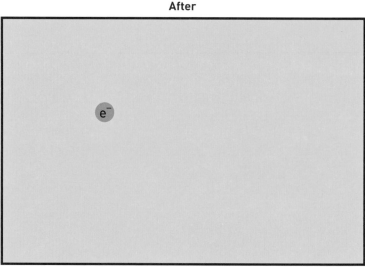

After the annihilation of one billion matter–antimatter pairs, there was one matter particle left over. The early universe was not exactly equal in matter and antimatter particles.

Condition 1 → The reactions such as particle decay and particle production have to inherently be able to produce more baryons than antibaryons.

Condition 2 → *C-symmetry and CP-symmetry* have to be violated. C-symmetry is the statement that if you have a reaction in which positive-charged particles produce negative-charged particles, there must also be interactions in which negative-charged particles produce positive-charged particles. Because matter and antimatter have opposite charges, interactions that produce more baryons than antibaryons must not be counteracted by interactions that produce more antibaryons than baryons (by swapping their charge values from C = positive to C = negative), otherwise the net baryon number would remain exactly zero. CP-symmetry is the statement that if both the charge (C) and handedness (P) of particles in a process are reversed, you still get an allowed process. CP-invariance (*charge parity invariance*) has to be violated to produce more baryons than anti-baryons because the 'handedness' or parity of baryons and antibaryons cannot cancel out.

Condition 3 → Interactions must be out of equilibrium. This means that reactions that create baryons cannot be exactly balanced by reactions that create antibaryons. This condition is easily met by the Big Bang expansion because it is constantly making the universe cooler and less dense as time elapses. This causes some reactions such as pair-production to experience

THE SAKHAROV CONDITIONS

Condition 1 ▶ *Baryon number violation*

Condition 2 ▶ *C-symmetry and CP-symmetry violation*

Condition 3 ▶ *Thermal non-equilibrium*

energy thresholds where pairs can decay but cannot be recreated. For example, a photon with an energy of about 1 MeV is needed to pair-produce an electron and an antielectron. These particles then combine and annihilate, but when the temperature falls below 1 MeV, these pairs can no longer be produced, so the production-annihilation equilibrium for these particles is lost as the universe expands and cools.

Andrei Sakharov

Commonly known as the Father of the Soviet Hydrogen Bomb, Sakharov was born in 1921 and as a nuclear physicist worked on fusion devices in the late-1940s. In 1950 he proposed the tokamak concept for controlled fusion, returning to his first love in science – cosmology – by 1965. He was especially fascinated by Big Bang physics and the problem of the matter-antimatter asymmetry in the modern universe, leading to his publication of the so-called Sakharov Conditions in 1967. This also led him to consider exotic cosmologies in which the CPT Theorem was satisfied by proposing two 'sheet' universes joined at the Big Bang singularity, in which one universe was matter dominated and the other antimatter-dominated.

For Condition 1, the Standard Model has no processes that change baryon number. Protons are the most stable of all baryons, and so baryon number change would have to imply the decay of protons. Protons do decay into neutrons, but since both particles are baryons, the baryon number remains the same before and after the decay. Currently, the lower limit to the half-life of protons has been measured to be upwards of 10^{34} years. The favoured decay would be into a neutral *pi* meson and an antielectron. The *pi* meson would then decay into two photons, so the final proton decay reaction involves an antielectron and two photons. The baryon number goes from one to zero, so baryon number is not conserved. There are, however, a number of mechanisms for proton decay offered in GUT and Supersymmetry Theory, but these have not been experimentally verified to date. Some unsuccessful versions of these theories can be eliminated because they actually cause proton decay to proceed faster than current upper limits allow.

For Condition 2, the violation of CP-symmetry was discovered in 1964 in the decay of neutral K mesons (kaons). There are two K^0 mesons with identical rest masses of 497 MeV. However, one type of K^0s, (called K^0-short) has a half-life of only 9×10^{-11} s while the other kaon, called K^0-long, has a half-life of 5×10^{-8} s. The result is that slightly more baryons are produced in the decays of these particles than antibaryons, but the rate is too small to account for the measured 1:1 billion cosmological imbalance.

For Condition 3, the out-of-equilibrium condition means that the universe expansion rate has to proceed faster than the decay time for the interaction so that the particle-antiparticle production that tries to keep the numbers equal can't keep up with the decreasing occurrence of their pair production as the universe cools. At least this condition can in principle be satisfied during the Big Bang so long as there is a threshold energy required for the process, above which equilibrium is temporarily achieved but below which there is no equilibrium to keep the abundance of matter equal to antimatter.

Currently there are no experimentally verified mechanisms within the Standard Model or Big Bang cosmology by the start of the Electroweak Era which satisfy the Sakharov Conditions and allow for more matter to be favoured over antimatter at the 1:1 billion level needed to account for our matter-dominated universe.

The Horizon Problem

During the early instants of the Big Bang, the universe was expanding at a tremendous rate, and this led to some severe issues with cosmological horizons and the uniformity of the matter and the cosmic background radiation during the expansion. During the Electroweak Era, when the age of the universe was only 4×10^{-15} s, from every point in space, you could only receive information from your neighbours if they were closer than 4×10^{-15} s $\times 3\times10^{10}$ cm/s = 0.0001 cm. Inside this horizon radius, given the estimated density at this time, there would only be about 18,000 kg (39,680 lbs) of matter. This is not enough to make anything interesting, but it is plenty large enough to have a bewildering number of elementary particles interacting at a ferocious rate with each other.

An artist's impression of a young galaxy, less than one billion years after the Big Bang. The rate of expansion of the early universe caused real difficulties with cosmological horizons.

THE SEARCH FOR COSMOLOGICAL ANTIMATTER

The realization that we live in a matter-dominated universe with little or no antimatter in equal measure has been an issue in cosmology since the discovery of antimatter by Paul Dirac in 1928, and the detection of the positron by Carl Anderson in his 1932 cosmic ray studies.

The contact between matter and antimatter produces a burst of gamma-ray photons. If there were any large reservoirs of antimatter within the Milky Way, this radiation should be easily detectable. All that is observed, however, is the small amount of antimatter produced from individual particle collisions in space.

Among the proposals for addressing the question of missing antimatter is that by some means matter and antimatter were segregated in space so that the nearest concentrations may be billions of light years from Earth and undetectable. There is no known physical mechanism that could perform this trick, so cosmologists have looked to other explanations.

Meanwhile, although the searches for the gamma-ray annihilation radiation have turned up no detectable emission, theoretical physicists such as Andrei Sakharov in the 1960s considered that processes near the Big Bang itself biased our universe in favour of normal matter. There could be mechanisms within the Standard Model or beyond which cause particles to decay in such a way that more antimatter is produced than normal matter.

Paul Dirac won the Noble Prize for Physics in 1933 for his work on atomic theory, but before this he had made perhaps an even more important discovery – the existence of antimatter.

Chapter Eleven
INFLATIONARY COSMOLOGY

Symmetries – The True Vacuum – The False Vacuum – Inflationary Big Bang Cosmology – Reheating and the Graceful Exit – Evidence for Inflation – Dark Energy

SYMMETRIES

In the last chapter, we began our journey to the Big Bang from the present time and arrived at the Electroweak Era – the current experimental horizon of research at a time some 4×10^{-15} s after the Big Bang. We also arrive at a time when our conventional understanding of the ingredients of the universe has begun to be challenged. We do not know where the CBR photons came from that dominate the matter particle content of the universe by over a billion to one. We do not know why a smidgeon of matter survived in what might have been a universe otherwise symmetric in matter and antimatter. We do not have an understanding of where or when dark matter particles arrived on the scene and became the dominant form of matter in the universe.

Dominance of matter

As for the Standard Model itself, there is no understanding of the many adjustable quantities in this theory. For example, what determines the number of generations of leptons and quarks, or what determines the way that the Higgs boson interacts with each kind of particle to give them their unique masses. If we wish to understand the reasons behind these cosmic circumstances, it seems we have no choice but to explore an even more remote landscape, largely populated by theoretical ideas with few indisputable data points to support them at this time.

The language of Symmetry we encountered in Chapter 7 entered cosmological investigations in the 1970s because this is the language used by physicists to describe the forces and particles needed to create the physical events and conditions of the early universe. Physicists see the transformations between forces and particles as a series of phase transitions, such as water vapour cooling to liquid at one temperature and then freezing to ice at a still-lower temperature. As the universe expanded and cooled, it is also viewed as having passed through a series of freezing episodes at specific temperatures (energies) in which the various symmetries among the forces and particles are successively broken as the universe relentlessly expanded and cooled.

Freezings

In today's universe, U(1), SU(2) and SU(3) are all broken symmetries because the three forces that they represent – the electromagnetic, the weak and the strong – have different strengths and behaviours.

U(1) ▶ *Electromagnetic force.*

SU (2) ▶ *Weak force*

SU (3) ▶ *Strong force*

But during the Electroweak Era, when the universe was heated to about 1 trillion K, the electromagnetic and weak forces became similar. In the new language, the universe became 'symmetric' for the electroweak force represented by the symmetry SU(2) × U(1), but it remained 'broken' for the electroweak force and the strong force represented by the symmetry SU(3). That's why at this time and earlier there were essentially two distinguishable forces represented by SU(2) × U(1) and SU(3). What was realized by the mid-1970s was that a symmetry group like SU(5) contained the symmetries for U(1), SU(2) and SU(3) and so it was now possible to envision a new progression of symmetry-breaking events in the early universe.

At incredibly high temperatures, the physics could be described by the full SU(5) symmetry in which the strong, weak and electromagnetic forces were completely unified. This situation continued until the universe expanded and cooled so that SU(5) broke into the separate SU(3) and SU(2) × U(1) symmetries so that the strong force became distinct from the electroweak force. This defined the start of the Electroweak Era. Then at 1 trillion K the electroweak symmetry of SU(2) × (U1) broke into the separate pieces of SU(2) and U(1) so the weak and electromagnetic forces now became distinguishable and the Electroweak Era came to an end.

THE TRUE VACUUM

When we think about empty space, we imagine a condition in which all elementary particles are removed and all free fields such as electromagnetic fields are also removed, leaving behind…well…nothing; except the ineffable fabric of three-dimensional space. General Relativity says that this three-dimensional Void is also a fiction because it is the embodiment

The Physical Vacuum of the Cosmos

According to Heisenberg's Uncertainty Principle (HUP), written as $\Delta E \Delta T \geq h/4\pi$, if you tried to state that the void is exactly empty, meaning that $\Delta E = 0$, then you could make this statement only if you observed this Void for an infinite amount of time so $\Delta T = \infty$. For any shorter duration, the Void would carry some latent energy due to virtual processes. For example, an electron–positron pair could come into existence with an energy $\Delta E = 2mc^2 = 1.2$ MeV. But so as not to violate HUP, it could only remain as a pair for $\Delta T = 2.7 \times 10^{-22}$ s. Any longer, and the existence of the pair would violate HUP, and you would be able to see the pairs form and evaporate, thereby violating the conservation of energy in front of your nose. In quantum field theory, these virtual processes can be incredibly complex, but all that is required is that their energy and duration not violate HUP. The entire rubric of quantum electrodynamics, quantum chromodynamics and weak interaction theory rest on the accuracy of HUP, and their predictions have been verified to a spectacular degree of accuracy. All is well so long as the virtual particle process remains 'scratch paper' in the cosmos. However, we can actually see the direct consequences of this virtual vacuum at work under laboratory conditions in processes such as the *Casimir Effect*, which is discussed in more detail later on (see page 200).

of the gravitational field. If you tried to remove the gravitational field threading through this void, you would only succeed in annihilating space itself. But even before we go to this extreme, quantum mechanics, and in particular quantum field theory, says that this Void is filled with the comings and goings of innumerable virtual particles, and literally things going bump in the night.

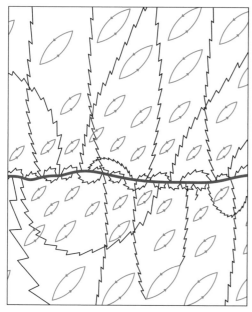

Empty space is filled with virtual processes that affect quantum systems such as particles and atoms.

THE FALSE VACUUM

Recall from Chapter 5 that in Grand Unification Theories (GUTs), the symmetry between the strong force and the electroweak force is broken by the action of a supermassive Higgs boson. This boson, like all other particles, is a quantum of its own field, just as photons are the quanta of the electromagnetic field. Electromagnetic fields always have a source, whether it is a charged electron or a charged atom, or a radio transmitter broadcasting the evening news. However, Higgs bosons are members of a class of particles that have zero quantum spin. When we represent these fields mathematically in the language of General Relativity, they are present throughout all of spacetime and not isolated to a particular source. In every cubic centimetre of space, there is also a piece of the Higgs field hiding away. Because it has the same properties everywhere in space, it interacts with matter in exactly the same way, whether the matter is on Earth or in the farthest galaxy we can see.

The energy of the vacuum is determined by the Higgs field causing the vacuum to be unstable.

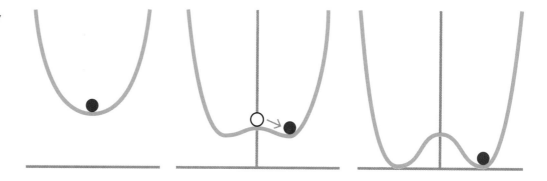

But Higgs fields, and in particular the quantum particles of which they are composed (Higgs bosons), can also interact with themselves. This self-interaction acts like potential energy, so that the more strongly they interact with each other, the greater will be the potential energy. By this mechanism, empty space can have normal virtual particles coming and going according to Heisenberg's Uncertainty Principle, but it can also have an average potential energy due to the underlying Higgs field. This Higgs vacuum energy is represented, mathematically, by what is called the Higgs Potential, and it has some interesting features that depend on the interaction energy. This interaction energy can be changed by increasing or decreasing the energy of collision of the other particles in the space; in other words, the temperature of the quantum system. The diagram on the previous page shows some representative Higgs potentials near the time of one version of the GUT symmetry-breaking event.

VACUUM ENERGY ▶ *the underlying energy of space that follows from Heisenberg's energy–time uncertainty principle.*

How the vacuum energy changes as it is heated.

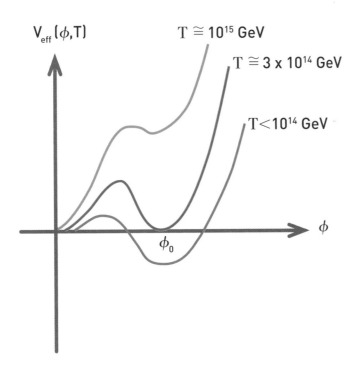

When the temperature (energy) is very high, the lowest point in the figure for the upper curve is at $\phi = 0$. Because the mass of the Higgs boson depends on the ϕ, this means that at very high temperatures above 10^{15} GeV, the Higgs boson has no mass. That means that the particles it interacts with also gain no mass, and so we are in the pure-symmetry state of SU(5). As the temperature falls to values below 3×10^{14} GeV, the Higgs boson acquires a mass equal to $\phi = \phi_o$. Upon further cooling, because there are no lower energy states for the Higgs field, it gets stuck at a value of ϕo. From then on, the physics is determined by this mass value for the Higgs, which determines how different the strong and electroweak forces will be. In actuality, things are more complicated than this, as we might expect. If this Higgs potential has any wiggles in it like the bump shown in the figure to the left, a very interesting thing happens as the system (universe) continues to cool.

The system can start out in the symmetric state at high energy near $\phi = 0$, but the Higgs field is able to change faster than the particles can keep up with it. As the Higgs settles into its lower-energy minimum at $\phi = \phi_o$, the particle system may still be trapped near $\phi = 0$. The diagram shows that the vacuum of the particle system is trapped in one minimum of the vacuum potential while the Higgs field is stable in the other minimum. They are separated by an energy barrier. Physicists say that the $\phi = \phi_o$ state is the true vacuum and $\phi = 0$ is the false vacuum. How does the particle system react to this?

We can copy its behaviour from models of radioactive decay where a nucleus suddenly ejects an alpha particle by a process called quantum tunnelling. The time it takes for the tunnelling through the energy barrier to occur depends on the height of the barrier energy and the energy of the particle.

Quantum Tunnelling

Quantum Picture

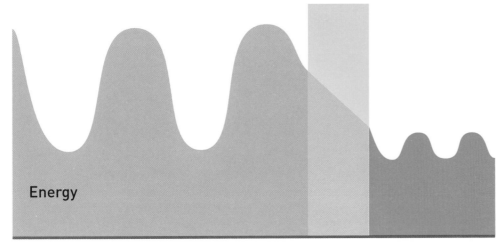

Energy

Inside Nucleus | Position Outside Nucleus

The alpha particle has a wavefunction inside the nucleus that is represented by the large-amplitude sine wave on the left side of the barrier. The distance above the horizontal line represents the energy of this alpha particle inside the nucleus. The height of the nuclear energy barrier is represented by the rectangle whose height is the barrier energy, and the width is an approximation to the thickness of the surface of this barrier at the nucleus. Because the barrier is not infinite in energy, the wavefunction of the alpha particle has to leak out of the nucleus, represented by the smaller sine-wave to the right of the barrier. Inside the barrier, the amplitude of the alpha particle wave function exponentially decays to a smaller level. The thinner the barrier, the less decay and the more-equal the inside and outside wavefunctions will be.

QUANTUM TUNNELLING ▶ *in quantum mechanics, when a particle such as an electron passes through a barrier that, in classical physics, should repel it. According to the Heisenberg Uncertainty Principle, the particle has some finite probability of overcoming the barrier, known as 'tunnelling'.*

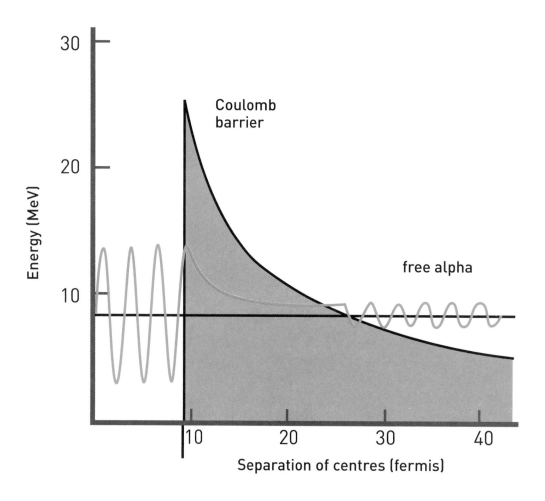

*An alpha particle can escape an atomic nucleus, even though it
does not have enough energy, by quantum mechanical tunnelling.*

The higher the energy difference between the two states, the longer it will take for the system to tunnel to the lower-energy state consisting of a separate nucleus and alpha particle. This is believed to apply to the Higgs–particle system as well. The particles in the *false vacuum* will eventually tunnel through the Higgs energy barrier and arrive in the *true vacuum*. The simplest way for

Alan Guth

This American theoretical physicist was born in 1947 in New Brunswick, New Jersey and received his PhD in physics at the Massachusetts Institute of Technology 1972 before accepting positions at Princeton, Columbia, Cornell and Stanford between 1972 and1979. While at Cornell University and Stanford University, he investigated the consequences of a phase transition in the universe just after the GUT phase ended. He discovered with Henry Tye that as the false vacuum transitioned to the true vacuum, space would expand exponentially and 'inflate' to enormous scales. This inflation would reduce the number of magnetic monopoles in the universe and also account for the 'flatness' of spacetime at the present time, as well as solving the Horizon Problem at the same time. For this, and other related work on inflationary cosmology, he was awarded, with Andrei Linde and Alexei Starobinski, the Kavli Prize in 2014.

Alan Guth developed the theory of cosmic inflation, which solved the Horizon problem.

this to occur involves the formation of bubbles of the true vacuum inside the false vacuum. The bubbles merge and this completes the transition from the false to the true vacuum. How does this apply to cosmology?

Between 1979 and 1981, physicist Alan Guth at Stanford University studied how this Higgs vacuum energy effect in GUT theory would apply to Big Bang cosmology, and he was greatly surprised by what he discovered. For technical reasons, it is believed that this vacuum energy is caused not by ordinary Higgs bosons but by another scalar field called the *'Inflaton Field'*.

INFLATON FIELD ▶ *a theoretical scalar field that may drive cosmic inflation in the early universe.*

Inflaton Field

INFLATIONARY BIG BANG COSMOLOGY

There is a direct relationship between the value of the *Inflaton Field*, ϕ, and Einstein's cosmological constant Λ. When $\phi = 0$, as in the true vacuum, the cosmological constant has a value of zero, and so the universe expands according to the normal Hubble expansion rate predicted by the Friedmann 'Big Bang' solutions. But when ϕ is not zero, the universe undergoes quantum tunnelling to the true vacuum. During this time, the cosmological constant will not be zero at all. This means that, instead of expanding linearly at the Hubble rate as the universe ages, it will expand at an exponential rate as predicted by the Einstein-de Sitter cosmological model, so that it will double in size with the passing of every time interval. This provides the explanation for why the CBR has the

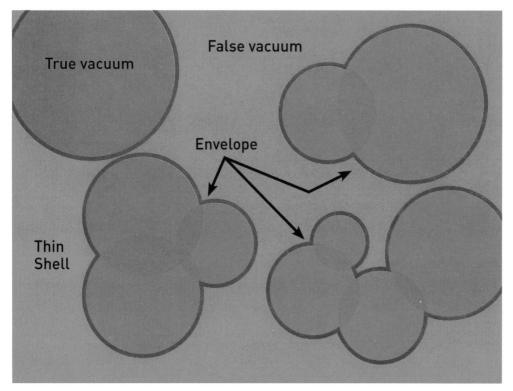

Quantum Tunnelling

The time the universe spends in the Inflationary Era doubling phase depends on just how fast the transition from the false to true vacuum is, and this depends on the potential energy difference between the two vacuum states.

The initial calculations by Guth showed that if the GUT SU(5) symmetry was starting to break near an energy of about $E(GUT)=10^{15}$ GeV at a time 10^{-35} s after the Big Bang, then the doubling period could be extended, depending on the exact shape of the Higgs Potential, to $E=10^{14}$ GeV after some 10^{-33} s. If particles started out at $0=10^{-33}$ cm apart (called the Planck scale), after only 100 e-foldings or 2^{144} doublings between 10^{-35} and 10^{-34} s, their separation would be $a(t)=10^{-33} \times 2^{144} = 105$ km. Note that by this time, 10^{-34} s, the horizon size is only 10^{-24} cm, which is far smaller than the scale of the matter that emerged from this one quantum patch after inflation.

An example of how the false vacuum changes to the true vacuum by forming merging bubbles.

As the sphere expands, the surface seems flatter and flatter with no curvature.

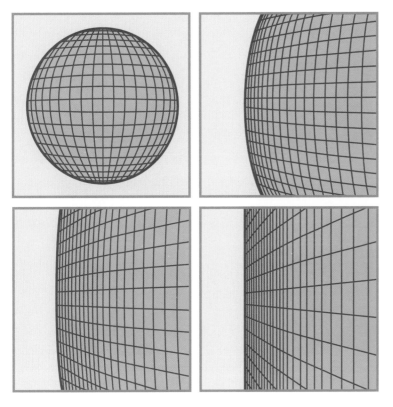

same temperature across the entire sky today. Everything we see around us emerged from one small quantum patch of spacetime just before inflation, and within which the temperature and physical properties of matter and radiation were very similar.

Once the tunnelling had completed, the ordinary Hubble expansion would resume as the vacuum energy reached a value of zero again. This period of doubling became known as 'cosmological inflation'. Its specific properties and duration are the subject of considerable study, not only because cosmological observations can help to define it, but because it also has immediate applications to theories that unify the forces of nature. Guth immediately saw how this mechanism solved two major problems in cosmology. The first of these was the 'flatness' problem. The second was the Horizon problem (see page 183).

Cosmological inflation

Flatness problem

The flatness problem has to do with the current cosmological models based upon observations that show our universe to have a very flat geometry in space(time). But how can it be so flat today? The answer from inflation is understandable by analogy to the surface of a sphere.

The surface looks very curved to start out, but if you enlarge the sphere to enormous size, its surface looks flatter and flatter as it evolves. This is what inflation does to an initially highly curved spacetime by dilating its scale enormously.

The second problem involves the temperature of the cosmic background today, which is very smooth to better then one part in 100,000. This cannot be possible from ordinary Big Bang cosmology because today, areas in the sky farther apart than about one degree should not have been in contact yet and should therefore show large differences from the nominal 2.7 K temperature.

What inflation does is to take a small region of the universe during what we can now think of as the 'Inflationary Era' and magnify this region by doubling until it is far vaster than the limits to our horizon today.

This patch could by now be thousands or millions of times bigger than our horizon distance of 14 billion light years, and so within our little patch we see a constant temperature. In other far more distant patches of spacetime, their current temperatures could be 3.2 K, 5.8 K or 0.5 K determined by the statistical variations that went on during the Inflationary Era.

REHEATING AND THE GRACEFUL EXIT

One problem with the original idea of *'inflationary cosmology'* was that there was no graceful way for this inflation to stop and then transition into the normal Hubble expansion predicted by Big Bang cosmology. Physicists such as Alex Vilenkin, Paul Steinhart and Andrei Linde offered new Inflationary Cosmology models with an added effect that seemed to solve the 'graceful exit' problem.

First, inflation is driven not by a relative of the ordinary electroweak Higgs field, but by a new 'Inflaton' field. This splitting is required because the predicted supermassive Higgs particles interact with matter so strongly that the universe would have imploded rather than expanded. A new field with a much weaker interaction with matter had to be proposed to cause *inflation*.

Next, in most versions of GUT and Supersymmetry Theory, there are predicted to be new families of very massive particles with masses near the GUT energy of 10^{15} GeV. The unique thing about these particles is that they contain features of both leptons and quarks, hence are often called *'leptoquark'* particles. Because of these 'mixed' properties, they can convert quarks into leptons and vice versa, thereby violating baryon number conservation. Over time, they can literally cause protons to decay along with the rest of the world around us.

Finally, the *quantum fluctuations* in the inflaton field gave rise to pair production of supermassive particle–antiparticle pairs. This seems to act like

Leptoquark particles

a break on the strength of the inflaton field, causing inflation to diminish in strength as it proceeded. As a result, the annihilation of these supermassive particle pairs also spawned the CBR and the major types of Standard Model particles along with reheating the universe, but not to the same energy as the GUT energy. The decay of these supermassive Higgs and leptoquark bosons was not symmetric, and it is at this time that the matter–antimatter asymmetry may have occurred.

Supermassive particle pairs

QUANTUM FLUCTUATIONS ▶ *following from Heisenberg's Uncertainty Principle, temporary changes in the amount of energy/ appearance of energetic particles out of nothing. It allows for particle-antiparticle pairs of virtual particles to form.*

Also, the quantum fluctuations in the inflaton field led to a spectrum of variations in mass density across each expanding patch, which became the seeds for the structure we see in the CMB anisotropy spectrum, and in the clustering of galaxies in the universe today. In fact, the spectrum seen in the CMB by Planck and WMAP matches almost exactly the prediction by inflationary cosmology and is considered a powerful vindication of the theory.

EVIDENCE FOR INFLATION

During the GUT Era prior to inflation, the density of matter at these scales and energies was subject to variations in strength dictated by quantum mechanics. As inflation proceeded, these quantum fluctuations in density were greatly enlarged, leading to a range of irregularities in the distribution of matter. The strength of these irregularities at different scales is predicted to follow a specific spectrum. This spectrum of variations will imprint itself on the cosmic background radiation, especially at large angular scales beyond 1 degree. Detailed studies of the Planck and WMAP data were able to confirm

Irregularities

by 2006 that the measured irregularities in the CMB have almost exactly the spectrum that is predicted by inflation. No other simple explanation currently exists for this observation. In fact, the inflationary spectrum has also been found in the clustering of galaxies measured by the Sloan Digital Sky Survey.

B-modes

Future tests of the inflationary phase in cosmic expansion will hopefully be obtained by the detection and measurement of 'B-Modes' in the polarization of the CMB. These are produced when the photons in the CBR scatter off irregularities in spacetime caused by gravitational radiation. Detecting them in the right amount would be strong evidence, not only for inflation but for gravitational radiation having been abundant before 10^{-34} seconds after the Big Bang. An initial confirmation of this polarization effect was presented in 2014 by the *BICEP2 Collaboration* (ten North American universities working together) but was deemed to be unconvincing proof a few years later.

DARK ENERGY

Discovery of dark energy'

We may not have to consider only the Inflationary Era to see this vacuum energy phenomenon at work. The discovery of dark energy in the data from WMAP and Planck, together with the detection of accelerated expansion discovered by the Type 1a supernova studies, suggests we are living through a new inflationary era today. The cause may either be a new wrinkle in the primordial inflaton field potential ϕ or it may be caused by yet another cosmological scalar field.

The Big Rip

The consequences of accelerated expansion today are rather severe. If it continues unchanged, it is estimated that, in 50 billion years, the Milky Way will be the only visible galaxy remaining in our neighbourhood. All of the other galaxies that now surround us will have been dragged by the accelerated expansion of space so that at their great distances they will be too faint to be observable. If the process continues beyond this point, in the far future, this space dilation effect will continue to strengthen so that our Milky Way will be ripped apart, then individual stars and planets, and finally even atomic systems will be dilated and shredded as spacetime itself continues to unravel. Astronomers call this dismal future the Big Rip. However, for this to happen, our current false vacuum state will have to persist for tens of billions of years, which would be an unimaginable degree of stability. Instead, it is more likely that this false vacuum state we live in today will decay to a true vacuum state as the new inflaton field finds its way to a still-lower energy level.

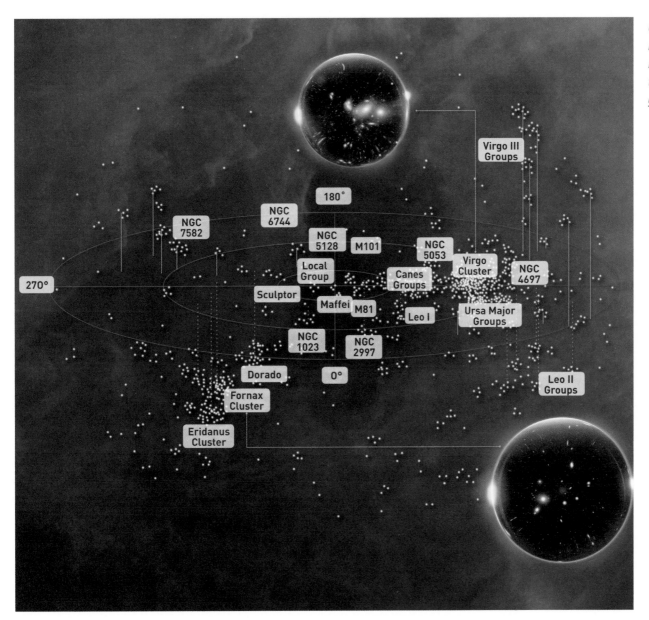

Quantum irregularities in the inflaton field became the seeds for modern galaxy clustering.

MEASURING THE VACUUM

When Heisenberg's Uncertainty Principle is combined with relativity, the uncertainty in measuring empty space to have exactly zero energy is replaced by virtual particles and processes that carry energy and mass for a brief period of time in a constantly fluctuating sea of hidden processes. The effects of these virtual processes are accurately calculated in the Standard Model, and have been directly measured in the properties of particles such as electrons. Within an atom, electrons experience a slight reduction in their energies causing what is called the Lamb Shift.

Another dramatic example of the existence of this virtual particle vacuum can be found in the Casimir Effect. Two parallel conducting plates are brought very close together, on the order of micrometres. The virtual vacuum outside these plates is normal. However, the vacuum between the plates blocks all of the virtual processes that have wavelengths comparable to the separation. That means there are fewer virtual processes going on between the plates than outside. This causes an attractive force between the plates that can be exactly calculated using the virtual vacuum model.

One interesting question is whether anything useful can be done with the Casimir Effect. Can you use it to do work and generate energy by 'robbing' energy from empty space? Because the Casimir force is repulsive and depends only on separation, it is classified as a conservative force. That means there is no path through time or space that can lead to net energy production. There have been many proposals for systems that take advantage of the vacuum's 'zero-point energy', but these are always defeated when the purported energy extraction process is looked at in detail.

ZERO-POINT ENERGY ▶ *in quantum mechanics, the lowest possible energy of a physical system at its ground state. It is still more energy than allowed for in classical physics because of the Uncertainty Principle.*

According to Alejandro Rodriguez at Princeton University, there is one possible application that does actually work. This involves the creation of low-friction nanomachines and gears. These devices are typically being fabricated at the scales at which the Casimir force is starting to become significant compared to frictional forces. The slight Casimir attraction can be tailored carefully to act as a repulsive force, and can then be used to overcome some of the normal friction between nano-scale parts, thereby increasing their efficiency.

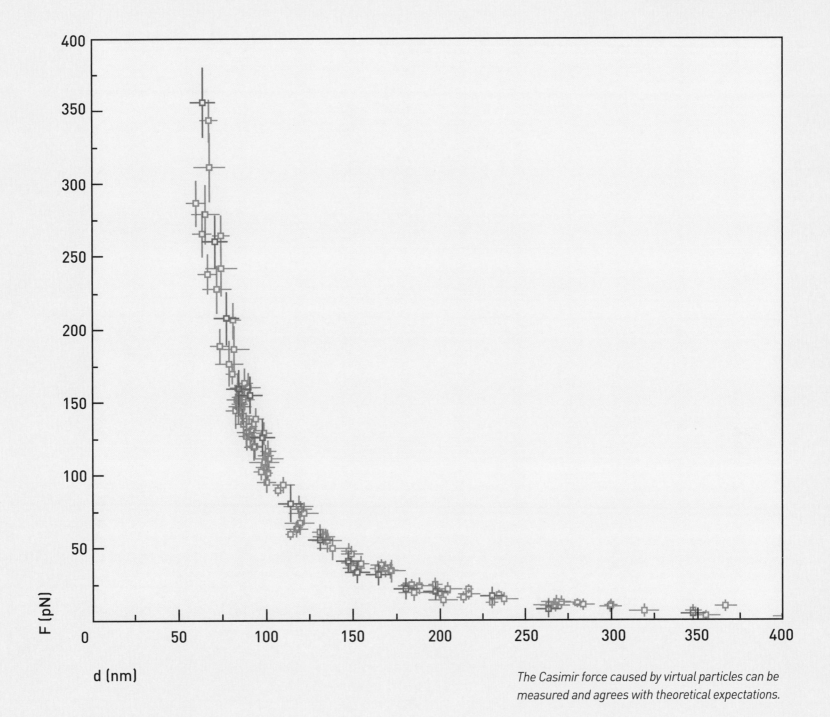

The Casimir force caused by virtual particles can be measured and agrees with theoretical expectations.

Chapter Twelve
COSMOGENESIS

The GUT Era – The Planck Era – Background Dependency and Independency – Loop Quantum Gravity – Structure of Spacetime in Today's Universe – String Theory – Eternal Inflation and the Multiverse v1.0 – Landscape and Multiverse v2.0 – Bubble Universes and the Multiverse v3.0

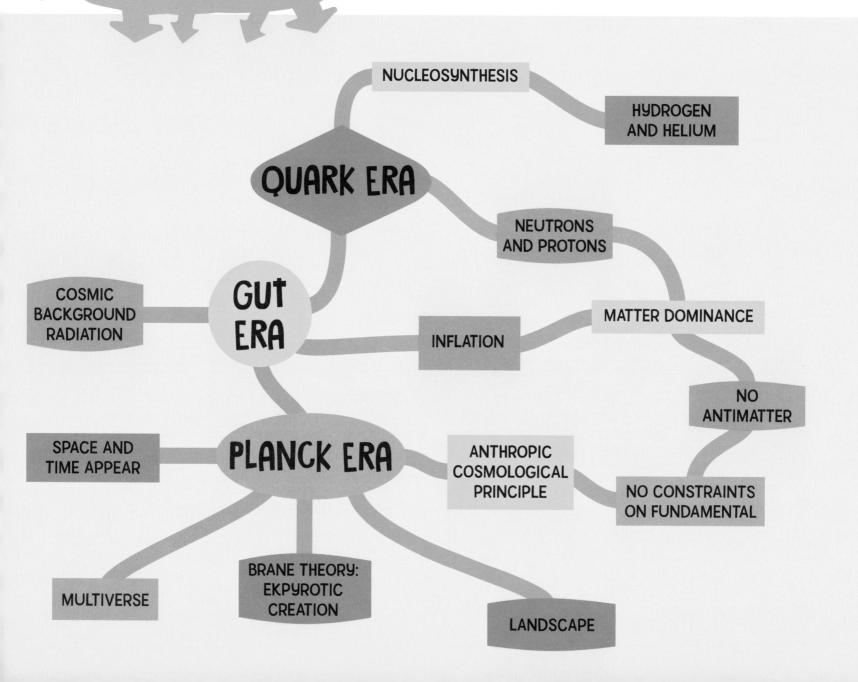

THE GRAND UNIFIED THEORY (GUT) ERA (10^{-37} SECONDS TO 10^{-43} SECONDS)

The four forces are merged

At 10^{-37} seconds after the Big Bang at the dawn of the Grand Unified Theory (GUT) Era, we conclude that the three interactions of the Standard Model – electromagnetic, weak and strong – merged together as one single force. What did the universe look like then? Spacetime was in the true vacuum state for GUT physics, so it was expanding at a constant rate just like the Hubble expansion after the Inflation Era ended.

Some cosmologists such as Roger Penrose have proposed in the 'Past Hypothesis' that this pre-inflation age had a very low entropy compared to the time after inflation was completed. The matter particles at that time consisted of supermassive particles of perhaps only a few varieties such as the X and Y bosons and the leptoquarks, which were all massless under the GUT symmetries, and with only one essential 'GUT' force, which represents the completely unified strong and electroweak forces. There may also have been far fewer CBR photons, making the photon-to-baryon ratio far smaller than the current 1:1 billion. Were there matter or dark matter particles present? Currently, theories predict such states, but only experiments can give us confidence that the underlying principles behind Supersymmetry and its competitors are on the right track. *Past Hypothesis*

Underlying the GUT Era is the mysterious nature of gravity and spacetime itself, whose energetic contortions produce a plentiful supply of free energy in the form of gravitational radiation from which to bring into being all of the remaining physics. Because of this, the next step in our description has to involve nothing less than the complete unification of all four fundamental forces, including gravity, into a single, logically consistent theory from which we can launch predictions about this intriguing and mysterious cosmological state. For this we need a theory that describes gravity and spacetime as a quantum phenomenon. *Unifying the four forces*

We do not know what this Mother of All Theories will look like, or what kinds of mathematics it will require, but physicists feel that there are many clues scattered about and within the descriptions we already have. The gravitational field (spacetime) has to start out with a grain, quantized description and then end up looking like ordinary general relativity. The shape of spacetime will also have to be the average over innumerable quantum states for its geometry.

At what scale is it expected that such effects will be important to cosmology? It is generally expected that, at the Planck scale, a new quantum-mechanical description for gravity and spacetime must be provided.

Planck Units

A key feature of combining General Relativity with quantum mechanics is that the gravitational field will be *quantized* (subdivided into small but measurable units or 'quanta'). It is generally recognized that this will happen at physical scales given by the Planck Units of mass, energy, time and space originally proposed as 'natural units' for physics in 1899 by Max Planck. These units can be formed by using the Newtonian constant of gravity (G), the speed of light (c) and Planck's constant (h) in the appropriate combinations to create the appropriate physical units as follows.

$$Lp = \sqrt{\frac{hG}{2\pi c^3}} \quad mp = \sqrt{\frac{hc}{2\pi G}} \quad tp = \sqrt{\frac{hG}{2\pi c^5}} \quad Tp = \sqrt{\frac{hc^5}{2\pi G k^2}}$$

When the appropriate values are used for these constants you get:
- Planck length $L_p = 1.6 \times 10^{-33}$ cm
- Planck mass $m_p = 2.2 \times 10^{-5}$ g
- Planck time $t_p = 5.4 \times 10^{-44}$ s
- Planck temperature $T_p = 1.4 \times 10^{32}$ K
- Planck energy $Ep = mp\, c^2 = 2.0 \times 10^{16}$ ergs or 1.3×10^{19} GeV

Compare these units to the proposed GUT Era scales of 10^{15} GeV and 10^{-37} seconds just before the Inflationary Era begins, and it is pretty clear that by the GUT Era, we are very close to the scale at which the gravitational field itself displays quantum properties.

The GUT Era is considered to exist between the Inflation Era at 10^{-37} s and about 10^{-43} s after the Big Bang, with the Planck Era occurring for times earlier than 10^{-43} s, if time itself is a meaningful concept. The difference in scale between the Planck scale at 10^{-33} cm and the GUT scale at 10^{-27} cm is a factor of one million. That means during most of the GUT Era, spacetime is still relatively smooth, like looking at a crumpled piece of paper from a distance of several hundred kilometres. But as we approach the Planck scale, we see more of the details of the 'crumpled paper' and these details become an ever more serious problem for describing the dynamics of the Standard Model particles. According to General Relativity, these curvature changes in spacetime represent energy, which is available to the elementary particles and fields as almost literally free energy. The geometric contortions of empty space itself is now a source of energy for creating whatever stable particle states the physical theory will allow. To accurately understand how these quantum curvature fluctuations will affect quantum fields, we need a fully quantum theory of spacetime. Since the 1930s, this has been something of a Holy Grail for physicists.

What this means for Big Bang cosmology has been a matter of deep and prolonged speculation driven by the search for a completely quantum theory of gravity.

John Wheeler envisioned the Planck Scale as a quantum foam of spacetime contortions.

THE PLANCK ERA (10^{-43} SECONDS AND BEYOND)

The physical properties of this state are largely unknowable. Normally, when we measure the properties and states of an object, we can choose to use photons to interact with the object. From the returned light properties we can infer the state of the object. In quantum mechanics, it is known that this process of observation interferes with the state and properties of the object being studied, and a limit is set by Heisenberg's Uncertainty Principle as to how detailed the information can be. One of the chief problems is that the wavelength of light sets the scale for how small the details are that can be seen or measured – a familiar problem for scientists using ordinary light microscopes. However, the finer the detail you require to measure, the shorter will have to be the light wavelength. Because the energy of the light increases as its wavelength decreases, you can measure accurately the properties of a quantum particle only by using the highest-available light energy.

For example, when you use a high-energy photon to measure the accurate position of an electron, the photon–electron interaction disturbs the motion of the electron in a way that cannot be corrected, and this is what results in Heisenberg's Uncertainty Principle. Quantum physics sets a limit to how well we can simultaneously know both the location and position of any quantum particle, each with perfect accuracy. If the cosmic gravitational field we also call spacetime follows similar quantum rules, the situation would be theoretically workable just as it is for ordinary quantum mechanics, but it seems this is not the case. When spacetime becomes quantized, we cannot perform even this simple measurement. As things stand, the Planck scale sets a limit to what we can ever know about space and time.

The photon we need to use to explore the quantum states near the Planck scale has to carry an energy of 10^{19} GeV in order that its wavelength of 10^{-33} cm is small enough to see these details. But a photon carrying that much energy immediately turns into a quantum black hole with a mass of 10^{-5} g. In 1975, physicist Stephen Hawking proposed that black holes can evaporate by a quantum mechanical process, emitting a steady stream of electrons, anti-electrons and photons, causing the black hole to lose mass. A quantum black hole evaporates by the Hawking mechanism after 10^{-43} s and the information we gather from the original photon is completely scrambled and randomized. This possibility was first proposed by the Russian physicist Matvei Bronstein in the 1930s. More recently, Carlo Rovelli at the University of Pittsburg has claimed to have shown that, at the Planck scale, dynamical systems cannot be described as evolving in terms of the universal time quantity we often denote by the symbol 't'.

There is no absolute time in terms of which an equation could be consistently defined to describe the evolution of any quantum state. When you reach the Planck scale at 10^{-33} cm and 10^{-43} seconds, you run out of test particles and clocks that are smaller than the phenomena for which you are attempting to formulate quantum laws. This is like trying to cut a 3 mm ($1/8$ in)

A possible glimpse of the Planck Era developed by Loop Quantum Gravity theory.

hole with a 5 cm (2 in) chisel, or like trying to feel beach sand through a winter glove. Curiously, many of the insights related to the ancient Planck Era also apply directly to the deep structure of spacetime in the universe today.

Gravitons

If quantum gravity at the presumed Planck scale of physics has the same type of mathematics as ordinary quantum mechanics, then just as the quantum field of electrodynamics is the electromagnetic field, the quantum field of gravity is spacetime itself. By analogy, just as the quanta of the electromagnetic field are called photons, the quanta of the gravitational field are called *gravitons*. Unlike photons, which have a quantum spin of 1, gravitons are also bosons but have a quantum spin of 2, and are also thought of as massless particles travelling at the speed of light. We have to treat the current state of spacetime as the superposition of many possible quantum geometries for spacetime coexisting in a blended state. The geometry for spacetime at any scale is then the average of many different states of spacetime geometry.

GRAVITONS ▶ *the quanta (discrete units or building blocks) of the gravitational field.*

Quantum spacetime

Because quantum spacetime geometry can be highly curved, and the three-dimensional shape of space curvature at a scale of 10^{-33} cm changes rapidly in a time of 10^{-43} s, these curvature changes can travel through spacetime as gravity waves defined by moving clouds of gravitons, and represent energy changes akin to $E=mc^2$. These high-energy gravitons would decay into a variety of particle states, if they exist, with energies between 10^{19} GeV and the GUT scale at 10^{15} GeV. In essence, we discover what earlier physicists had proposed, that the true nature of matter is that they are features of the geometry of spacetime itself. Here,

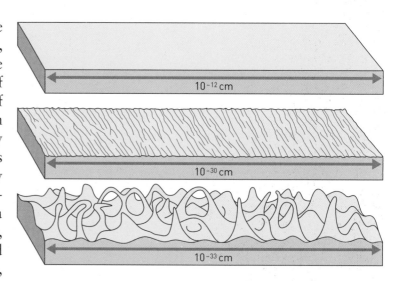

According to John Wheeler, at large scales, spacetime may be smooth but as we approach the Planck scale, the fluctuating quantum nature of spacetime emerges.

during the Planck Era, we see this unity in full expression with curvature in spacetime producing particles, and particles in turn producing curvature.

At the quantum gravity scale, we can also think of spacetime as consisting of innumerable patches that have differing temperatures and particle content. These patches eventually begin expansion until we reach the end of the GUT Era, at which point their scales increase exponentially as the Inflationary Era commences. After inflation, we are left with innumerable patches where the properties are still constant within them, but which differ statistically from one to another. The current observable universe is one small spot within one of these enormous patches, and this is why the CMB has virtually the same temperature in all directions. Depending on how long the Inflationary Era lasted, it is predicted that the patch size before inflation may have been only 10^{-26} m across, but after inflation it was 10^{24} m across, and by today it might be 10^{28} million light years across. That would be 10^{25} times farther from us than the most distant quasar. Our visible universe is literally a 'dot' in a far-vaster cosmos that, statistically, may nevertheless still resemble what we see around us today.

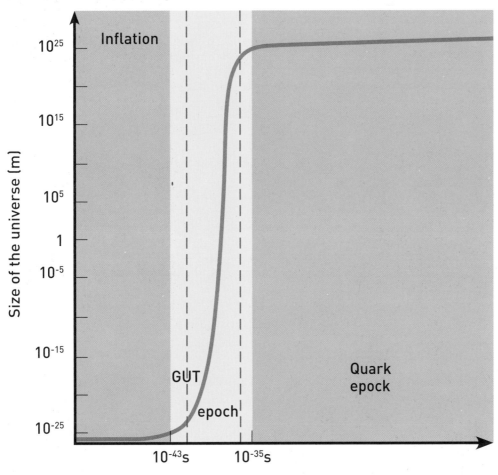

A calculation of the change in scale of the cosmos before and after inflation.

According to inflationary cosmology, the distances between the current vast patches that emerged from the Big Bang are still increasing exponentially in the false vacuum, so there will be no opportunity to ever see them, no matter how old the universe becomes. We may only hope to see to the limits of the current patch we live in, and within which the visible universe horizon is only expanding at light-speed.

One way to envision the Planck Era is as a 3-D foam of space with complex connectivity.

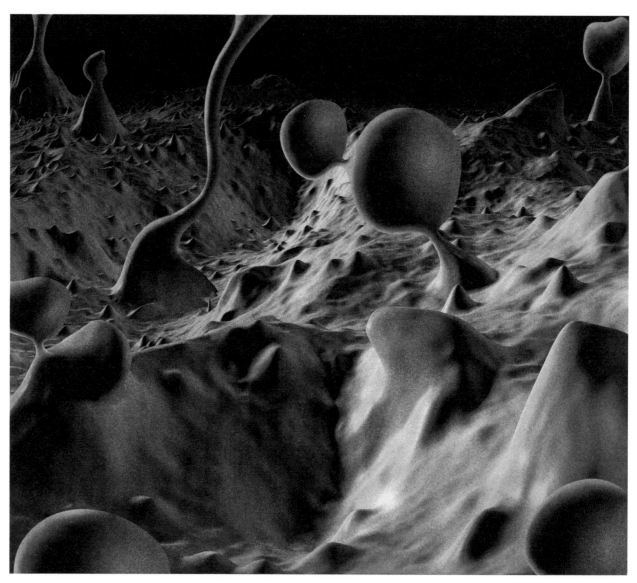

The above description of quantum gravity as applied to cosmology is only a rough approximation of the differing ideas which are emerging from the search for a unified quantum gravity theory that describes the Standard Model, and incorporates General Relativity and gravity. These ideas must account for some of the puzzling anomalies that continue to turn up in physics and cosmology, including dark matter, dark energy, the matter-antimatter imbalance, and the details of inflationary cosmology. There are two major approaches that have emerged over the last 50 years: Loop Quantum Gravity (LQG) and Superstring Theory. But first, let's explore the deep structure of these ideas.

Gottfried Wilhelm Leibniz suggested that space and time arose from the interaction between bodies.

BACKGROUND DEPENDENCY AND INDEPENDENCY

As we learned in Chapter 3, Sir Isaac Newton believed there was a fixed, absolute space and time that stood aloof from the movement of bodies, and to which one could ascribe various co-ordinate systems to define space and time locations. Leibniz, on the other hand, proposed that space and time have no fixed definitions but are features of the interactions between bodies. He said that the idea of space and time is in some sense actually created out of relationships between bodies and does not pre-exist. The former point of view is the basis for classical 'Newtonian' physics; the latter opposing point of view became the perspective for relativists and led to Einstein's relativity. But Newtonian physics was far easier to develop, mathematically, using the absolute space and time framework than relativistic spacetime, so Newtonian physics developed quickly and dominated the scientific world until the early decades of the 1900s.

Educational Psychology in Action

When Einstein developed Special Relativity in 1905 and General Relativity in 1915, his perspective was that Newtonian space and time as absolute properties of the world do not exist. Everything is observer–dependent. Moreover, the idea of space itself is a fiction. Instead, the only things that exist are objects and, as Leibniz asserted, space and time are derivable from relationships between these objects with no other ingredient required. The success of Einsteinian relativity through extensive experimental testing supports this foundational idea about space and time.

Worldlines

The relativistic concept of space and time, embodied in the concept of spacetime proposed by Hermann Minkowski, says that bodies move along worldlines, and it is the relationship between these worldlines that gives us the experience of space and time as real features of the universe. We can imagine these worldlines embedded in some larger, continuous four-dimensional geometry (mathematicians call this a *manifold*) like a line drawn on a two-dimensional sheet of paper, but relativity does not require us to do so. Instead, it is only the intersections of these worldlines, called Events, that define unique co-ordinates. Relativistic spacetime is actually sparsely populated

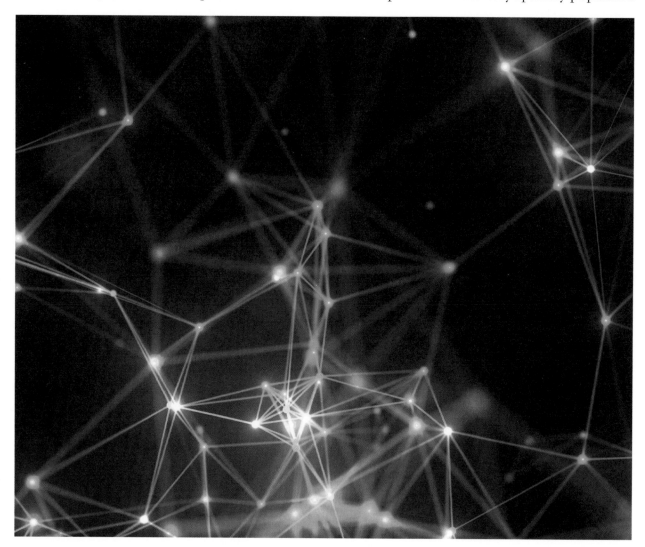

The network of events connected by worldlines creates what we think of as space, but space does not pre-exist.

by co-ordinate points for a finite number of interactions between bodies. This is unlike the mathematical construct for spacetime that is a manifold consisting of an infinitude of co-ordinate points. Nevertheless, from the physical spacetime created by the numerous intersecting worldlines, and the intrinsic geometry of these worldlines, we can mathematically derive the global geometric properties of the physical system in which worldline curvature relative to other worldlines is interpreted as a gravitational force.

We call this a background-independent framework because relativity does not require the a *priori* existence of space and time co-ordinate points in order to state the relationships between bodies and how they behave. Co-ordinate points in the background manifold that do not coincide with physical events do not have any effect upon the physics of these bodies. For calculations involving gravity, the mathematics that derives from this relativistic viewpoint is absolutely necessary for understanding many exotic gravitational phenomena. However, there is one major theoretical system that does depend upon a pre-existing background spacetime: quantum mechanics.

Every quantum system and its mathematical description requires a set of fixed co-ordinates in space and time. These co-ordinates come from an observer-specified reference system and are essential for defining a particle's state (the sum of all possible states for a system is called its wave function). The comprehensive Standard Model is actually a semi-relativistic statement of how particles behave, which borrows the principles of Special Relativity, and from which the energy-mass equivalence principle, $E=mc^2$, and other tools of this theory are used to create relativistic quantum field theory. But because gravity plays no role in defining a particle's quantum state, the Standard Model operates as though spacetime is completely flat and merely serves as scaffolding that facilitates calculations. The Standard Model is a background-dependent theory because it requires a pre-existing spacetime in order to formulate the properties of particles and their interactions.

The contrast between the background-independent theory of General Relativity, and the background-dependent quantum mechanics of the Standard Model is the major reason why these two fundamental theories of the physical world have not been brought into a more unified scheme. We will now look at two recent approaches that offer some prospects of meeting this unification challenge.

The electron clouds within atoms are only defined by their 3-D spatial locations.

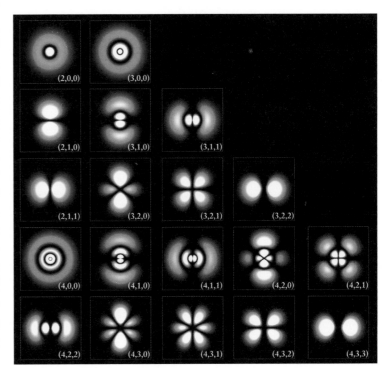

213

LOOP QUANTUM GRAVITY

The basis for loop quantum gravity (LQG) is to assume that:
- the Standard Model is correct and is an important data point;
- the proper theory describing space, time and matter is a background-independent theory following relativistic principles;
- the scale at which the theory must operate is near the Planck scale; *and*
- there are no extra dimensions to spacetime beyond the four we already know.

In the 1990s, two theoretical physicists, the American Lee Smolin and the Italian Carlo Rovelli, created a new way of looking at the unification problem which continued the efforts of John Wheeler and Roger Penrose in the 1960s. They took a long hard look at the foundation of spacetime within the relational perspective of Leibniz and Einstein and followed the mathematics to what seemed to be its logical conclusion. Spacetime has to be built up from something that is even more primitive than space and time.

An example of a spin network. Only the points represent space. The lines represent non-space relationships that carry area. The numbers represent the J-values in the equation for A_J.

Lee Smolin

Lee Smolin is an American theoretical physicist born in New York City in 1955, who received his PhD in physics from Harvard University in 1979. Almost from the beginning of his research career, he was compelled to investigate issues in quantum gravity theory, with the particular perspective that string theory was not the correct approach due to its 'background dependence' on a pre-existing spacetime. Together with Carlo Rovelli, Ted Jacobson and Abhay Ashtekar, he developed a completely new formulation of quantum gravity based upon ideas proposed in the 1970s by Roger Penrose. Since the announcement of 'Loop Quantum Gravity in 1994, he has been an outspoken critic of string theory and an innovative developer of observational tests of quantum gravity.

STRUCTURE OF SPACETIME IN TODAY'S UNIVERSE

At the Planck scale, spacetime becomes resolved into minuscule volumes that, according to the relational perspective, are connected to each other by relationships that then define how much area these points contain. They used as their analogue the spin network approach developed by Roger Penrose in his complex Twistor theory. Each vertex in the network represented a quantum of space volume at the Planck scale. The relationships between these volumes were drawn as a set of lines connecting them. Each line represented an elementary quantum of area (A_J) whose magnitude was derived from an index (J) according to the quantization formula:

Twistor theory

$$A_J = 8\pi l^2 \sqrt{J(J+1)}$$

where l is the Planck length (1.6×10^{-33} cm – see pages 204–5).

The amount of area associated with the volume was then just the sum of the area elements carried by each of the lines connected to the vertex volume. The total volume of space was also just the sum of the number of vertices. Just as the energy levels within an atom are quantized in terms of an energy index, so too is the area associated with this quantum theory for space, represented by the symbol J in the formula for A_J.

If you were to trace a loop through one of these spin networks that began and ended on the same vertex volume, you would find that this loop satisfies an important equation in Quantum Gravity Theory called the Wheeler-DeWitt equation, developed in the 1960s. These loop solutions are what give this quantum theory its name because, like field lines around a magnet (which show the direction of a magnet's force), these loops are critical to weaving together the quantum units of spacetime into what we experience as an integrated whole in terms of areas and volumes for space. This is a young theory, so much more work is needed to connect this description of gravity and spacetime with the Standard Model. Apart from gravitons, there is no current way to go from Loop Quantum Gravity (LQG) to actual Standard Model particles and fields. LQG is only a theory of spacetime and gravity. Also, it is strictly a theory of four-dimensional spacetime, and it has yet to be shown how the equations of LQG lead to equations resembling Einstein's equations for gravity in General Relativity at much larger scales.

Wheeler-DeWitt equation

STRING THEORY

String Theory was developed in the early 1980s at about the same time that work on Supersymmetry was accelerating. However, although Supersymmetry is a concept independent of String Theory, String Theory largely depends on the existence of Supersymmetry to make contact with the Standard Model world we live in.

Vibrating loop

In 1982, John Schwarz and Michael Green developed a mathematical approach to fundamental particles in which a particle was replaced by a vibrating, one-dimensional loop. What this loop physically represented was not important, only that a particle's internal structure could be mathematically specified in this way. That said, a particle would now travel through spacetime as a vibrating tube rather than a mathematical point mass whose worldline was a simple line. This alteration in representing a particle's structure led to further developments. One of the biggest was that you couldn't create a quantum theory for these string particles unless you greatly enlarged the dimensionality of spacetime to ten dimensions. Since four of these dimensions referred to our normal three-dimensional space and time, these dimensions had to remain very large and potentially infinite, but the other six had to be smaller than the atomic scale. The size of these dimensions was related to the tension in the string and, for a variety of simplifying reasons, it was generally considered that these dimensions were finite and of the order of the Planck scale in length. This was, after all, the only natural limiting scale offered by the universe for small things.

Calabi-Yau spaces

It was soon also realized that these six 'compact' dimensions could form their own geometric configurations called 'Calabi-Yau' spaces. Each of these configurations had their own particular symmetries, and these symmetries were directly related to the number and properties of the fermions and bosons that they could describe. For example, the number of topological holes in these geometries (called the genus) was found to be related to the number of families of fundamental particles in the Standard Model.

M-theory

Following considerable effort in the 1980s and 1990s, other researchers identified a total of five separate formulations of String Theory in ten dimensions that incorporated Supersymmetry, so 'Supersymmetric String Theory' offered not just one unified theory but five from which to choose. In 1993, physicist Edward Witten discovered that these five string theories could be united into one theory, called M-theory, which essentially described the

An artistic rendering of a quantum string.

unification process from five different perspectives. By adding one more dimension to spacetime for a total of 11, you once again end up with only one unifying theory.

What do these added dimensions really represent? We have a good intuition about the three dimensions of space and the one dimension of time, but these added dimensions are as different from normal spacetime, as the dimension of time is from one of the dimensions of space. As Nobel physicist Stephen Weinberg noted in the 1985 book *Perfect Symmetry* by Heinz Pagels: *"I talked about the extra 6 dimensions wrapping themselves up, but that's not necessarily the way one thinks about it now. One thinks about the theory formulated in four dimensions but with some extra variables which can, in some cases, be interpreted as coordinates of extra dimensions, but needn't be. In fact, in some cases, CANNOT be."*

This theory was further refined by a variety of theorists, including physicist Lisa Randall, who connected this theory with a cosmological setting. To do this, subsets of spacetime called 'branes' had to be considered. In the simplest case, our universe occupied a four-dimensional spacetime brane within this eleven-dimensional arena called 'the bulk'.

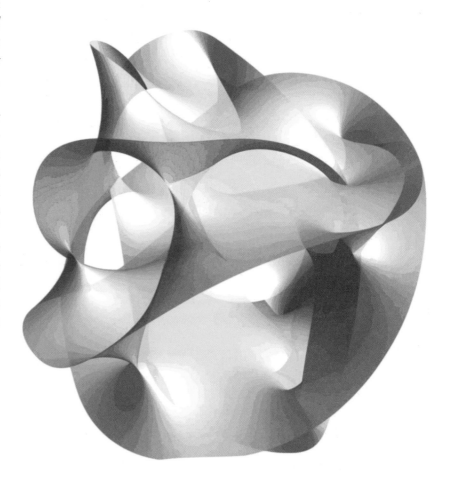

The additional dimensions to space may be extremely small and finite, defining their own geometries.

Lisa Randall

An American theoretical physicist born in Queens New York 1966, Lisa Randall went on to receive her PhD in physics at Harvard University in 1987 with Howard Georgi – one of the developers of SU(5) Grand Unification Theory. Randall became the first tenured physics professor at Princeton University and returned to Harvard University in 2001. She is active in the investigation of the consequences of M-Theory developed by physicist Ed Witten in the 1990s, and together with Raman Sundrum developed the Randall-Sundrum theory of 5-dimensional warped geometry that has led to cosmological investigations based on (3+1)-dimensional branes. She is also a prolific science popularizer and was named by *Time* magazine in 2007 as one of the 100 most influential people.

All of the Standard Model particles and fields exist only within this brane as strings with open ends rooted into the 'fabric' of 4-D spacetime. A particle has a very weak extension into the bulk by an amount limited by the Planck scale. Gravity, represented in String Theory by a closed loop, however, is free to travel throughout the bulk, which is why we experience gravity as such a weak force, cosmologically. The reason is that most of the strength of gravity is lost in vibrating among the higher dimensions of the bulk. Another interesting feature of Brane Cosmology is that it provides a new perspective on the Big Bang event itself. Called the Ekpyrotic Big Bang, these brane universes feel a weak gravitational pull between themselves. If they were to collide, they would convert huge amounts of energy into particles and fields throughout all spacetime, thereby creating a Big Bang event.

Our 4-D spacetime exists within an 11-D spacetime called 'the bulk'.

'Branes are essentially membranes – lower-dimensional objects in a higher-dimensional space…think of a shower curtain… not everything need travel in the extra dimensions even if those dimensions exist. Particles confined to the brane would have momentum and motion only along the brane, like water spots on the surface of your shower curtain.'

LISA RANDALL, PROFESSOR OF PHYSICS AT HARVARD UNIVERSITY.

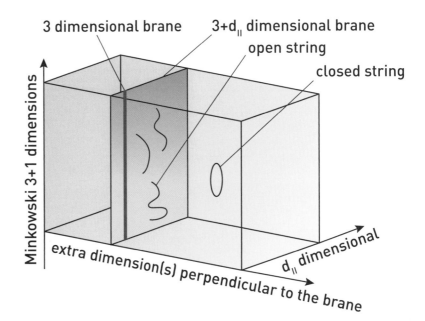

At the present time, String Theory and LQG have collaborated in a variety of calculations related to black hole physics, but while String Theory provides a possible means for accounting for aspects of the Standard Model but not gravity, LQG provides a detailed description of gravity and spacetime but not the Standard Model or other extensions to it. Meanwhile, we have a variety of new ways to describe the earliest conceivable events in the history of our universe near the Planck Era.

Lisa Randall helped to develop one form of Grand Unification Theory.

ETERNAL INFLATION AND THE MULTIVERSE V1.0

In Guth's original version of inflation, the transition from a false vacuum to a true vacuum was akin to bubbles forming within a carbonated soft drink. As the transition proceeded, these bubbles would expand, collide and merge to create our current true vacuum universe. The problem is that these collisions would create a very inhomogeneous universe unlike the relatively smooth and uniform one we actually live in.

According to the new inflation theory, developed in 1982 by Andrei Linde, Andreas Albrecht, Paul J. Steinhardt and Alex Vilenkin, the inflationary phase of the universe's expansion lasts forever throughout most of the universe. Because the regions expand exponentially rapidly, most of the volume of the universe at any given time is still inflating. Within this inflating bulk, individual bubble universes of true GUT vacuum emerge, but between them space is still expanding exponentially in the GUT false vacuum state. Eternal inflation, therefore, produces a hypothetically infinite universe, in which our visible universe is a vanishingly small patch within one of these bubble universes.

Data on the CMB provided by the Planck satellite suggests that the simplest textbook inflationary models are eliminated, but also imply that the starting conditions for inflation are far more complex and may lead to less inflation. As a result, the idea that inflation is eternal seems not to be vindicated, and studies of matter within individual bubble universes suggests they are filled with singularities (black holes). Therefore the existence of a multiverse of separate bubble universes no longer seems supportable within inflationary cosmology. But String Theory still seems to require such a possibility.

Bubbles forming in water as an analogue to vacuum phase change.

THE LANDSCAPE AND THE MULTIVERSE V2.0

String Theory combined with Supersymmetry led to the development of Superstring Theory. It is the next step beyond the development of purely supersymmetric extensions of the standard model such as MSSM described in Chapter 5. It describes physics that spans the entire Particle Desert between 100 TeV and 10^{15} GeV, and up to the Planck scale of energy at 10^{19} GeV. However, when specific calculations are attempted to tie down the number of particle states and their interactions, the plethora of possible Calabi-Yau spaces becomes a problem. By some calculations, there are 10^{500} different ways these spaces can be wrapped up, implying at least as many different solutions for what Standard Models should look like in our four-dimensional world. This collection of possibilities is called the *'Landscape'*. Presumably at, or near, the Big Bang, our universe had to 'select' which of these Landscape universes it wanted to manifest. Currently there is no known way for this selection to be made naturally, and so because of the statistical nature of the theory, one popular solution for now is to allow all of these possibilities to exist in a Multiverse of possible universes.

BUBBLE UNIVERSES AND THE MULTIVERSE V3.0

Other approaches to cosmogenesis that involve bubble universes start with the singularities within our universe called black holes. According to some theories developed in the 1980s and 1990s by Leonard Susskind and others, at the quantum gravity scale, pieces of spacetime are continuously being formed. They then pinch off from our spacetime to form separate 'bubble' universes. Most of these, statistically, remain Planck-scale universes that eventually vanish. However, others may meet some kind of threshold condition and expand to become large universes like our own.

In other versions of this idea, within black holes, the singularity actually represents such a bubble universe pinching off from our own. It leaves behind on our side an *event horizon* (the boundary from which no radiation including light can escape). But within the bubble universe it rapidly disconnects from our spacetime to become its own Big Bang, inflating spacetime. There can be no communication between our universe and this new one because its spacetime is completely disconnected from ours through the appearance of the event horizon and the singular state, which cannot be penetrated from our side to extract information.

Event horizon

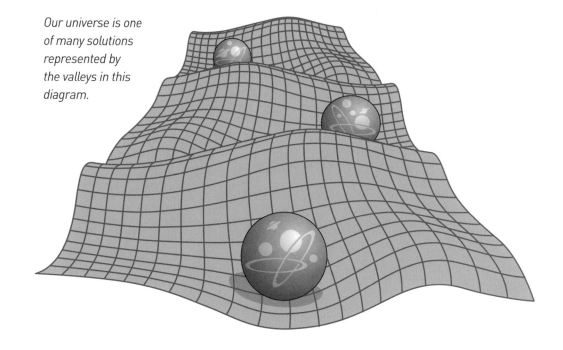

Our universe is one of many solutions represented by the valleys in this diagram.

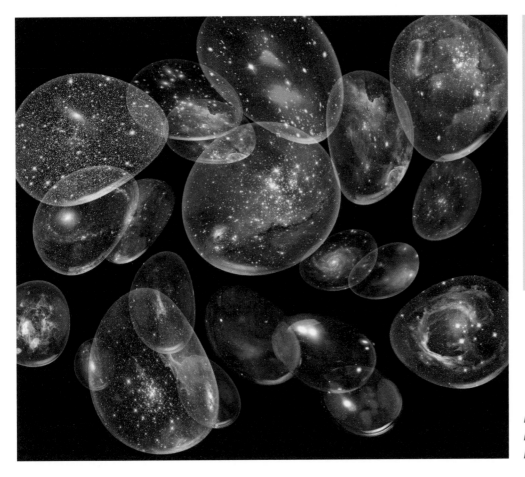

NAMES TO KNOW:
THE MULTIVERSE

Leonard Susskind *(1940–present)*

Alan Guth *(1947–present)*

Andrei Linde *(1948–present)*

Lee Smolin *(1955–present)*

Baby universes may be spawned from black holes.

According to Lee Smolin, this production of bubble universes leads to an interesting 'Darwinian' situation. He proposes that these new universes carry some information about the physical laws and the Standard Model from our universe. Because the Standard Model and our form of gravity favour the creation of stellar-mass and supermassive black holes, these new universes will also favour their creation, and so the numerous bubble universes spawned from our spacetime will in turn spawn their own vast collections of black holes and next-generation bubble universes also marked by our physical laws. Since our physical laws favour the creation of sentient life, the multiverse of Smolin bubble universes favouring life will eventually outnumber all of the 'failed' universes that do not by a process of cosmic Darwinism.

THE ANTHROPIC PRINCIPLE

String theory and inflationary cosmology seem to require more than one universe among a vast number of other possible universes. The reason for this is that these theories express themselves in terms of equations that have adjustable constants. For our universe, these constants are determined through observation, but the equations themselves offer the possibility of an infinitude of entirely equivalent solutions, each with their own set of adjustable constants. If each of these solutions applies to its own 'universe', then mathematically we have a multiverse of possible universe – one for each possible set of adjustable constants. The question becomes, how do we find ourselves in *this* particular universe?

Life as we know it, and specifically organic chemistry-based sentient life, requires the precise tuning of a handful of fundamental parameters in nature such as the value of the speed of light, Planck's constant and Newton's constant of gravity. Other quantities specify certain quantum mechanical relationships such as the number of Standard Model particles and fields, along with the values of the more than two dozen constants within the Standard Model which currently have to be assigned by direct observation. It has been a standing mystery how it is that, as John Wheeler reflects, 'the universe saw us coming from soon after the Big Bang'.

One solution to this fine-tuning problem involves what is called the Strong Anthropic Cosmological Principle. This states that there may, in fact, be an infinitude of separate universes existing in some manner of multiverse, and each of these represents a specific, but random, selection of fundamental constants and Standard Models. Most of these universes will be sterile of our kind of life and hostile to it. But others will be suitable for our kind of life, or other forms of life, with just the right constants and just the right number of particles and fields in their versions of the Standard Model. In most cases you just need to be within 10 per cent of the known values to make a viable universe. If the gravitational force is too strong, stars will rapidly evolve to white dwarfs before life can emerge on any orbiting planets. If Planck's constant is too large, atoms will be unstable and the chemistry for life will never arise.

However, these universes exist far beyond our visible universe in our four-dimensional spacetime, or are separated from us as other 'branes' within an eleven-dimensional 'bulk', and so are entirely unobservable. In fact, because our spacetime region is separated from theirs, there is no way to receive information about their existence. This means that they are essentially beyond the scientific method to study and to prove/disprove their very existence. There appears to be no way to test the Anthropic Principle that does not lead to circular arguments or experiments beyond scientific or technological execution within our universe.

The Strong Anthropic Cosmological Principle suggests that there may be an infinite number of universes.

Chapter Thirteen

THE FAR FUTURE

What Lies Ahead – The Near-Future – Star Death – The Big Rip – The Big Chill – Eternity – The End

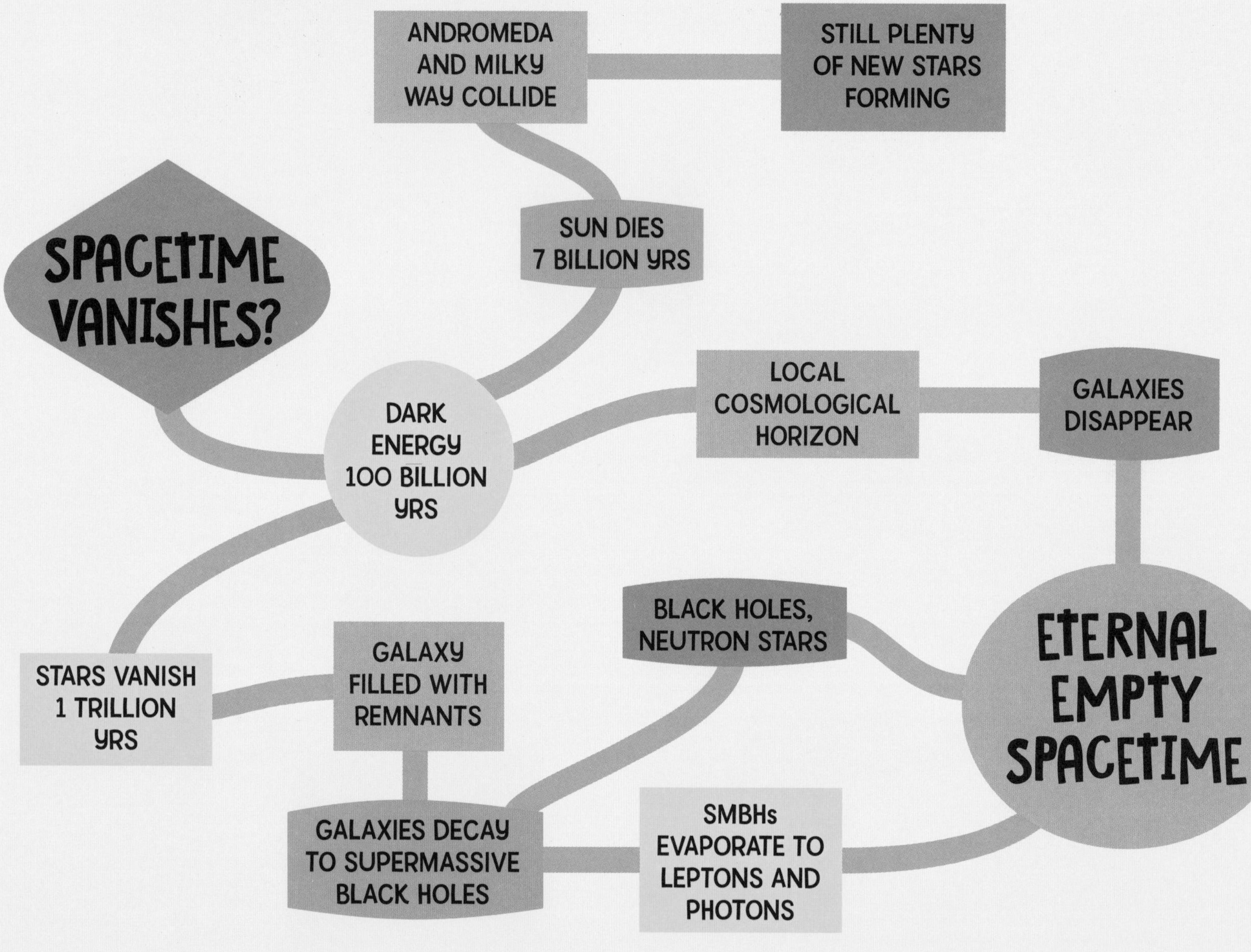
ANDROMEDA AND MILKY WAY COLLIDE
STILL PLENTY OF NEW STARS FORMING
SUN DIES 7 BILLION YRS
SPACETIME VANISHES?
DARK ENERGY 100 BILLION YRS
LOCAL COSMOLOGICAL HORIZON
GALAXIES DISAPPEAR
BLACK HOLES, NEUTRON STARS
ETERNAL EMPTY SPACETIME
STARS VANISH 1 TRILLION YRS
GALAXY FILLED WITH REMNANTS
GALAXIES DECAY TO SUPERMASSIVE BLACK HOLES
SMBHs EVAPORATE TO LEPTONS AND PHOTONS

WHAT LIES AHEAD

As physicists and astronomers worked diligently to discern the details of the origin and evolution of our universe up until the present time, other astrophysicists explored what the future might hold in store for this universe in the far future. At first these remote times were simply swallowed up in the description of the Friedman cosmologies. If the universe is flat or negatively curved, the universe will continue to expand for eternity, but if it is positively curved it is destined to re-collapse in a not-too-distant future perhaps less than 100 billion years from now. What has been ignored in the working out of these mathematical solutions is any details about the future evolution of galaxies, stars and even life itself.

In a benchmark paper published in 1977, Jamal Islam at the University College in Cardiff laid out the major events to come in a universe destined to expand for eternity. The basic idea offered was that small changes multiply to become major events after tens of billions of years, including the gravitational disruption of galaxies and the formation of supermassive black holes as systems lose gravitational energy by radiating gravity waves. Even low-probability events such as quantum tunnelling of macroscopic objects have time to complete their transformations over timescales measured in googleplexes of years (10^{100} years). In an independent but equally influential paper 'Time without end: Physics and biology in an open universe', physicist Freeman Dyson overlaid on the dismal and relentless dissolution of stars, galaxies and matter an optimistic outlook for organic life as the aeons pass. As he noted in his article: *'The general conclusion of the analysis is that an open universe need not evolve into a state of permanent quiescence. Life and communication can continue forever, utilizing a finite store of energy, if the assumed scaling laws are valid.'*

A simulation of the collision between Andromeda and the Milky Way.

THE NEAR FUTURE

Over the course of the next few billion years, our local universe is destined to change as red supergiant stars, such as Betelgeuse, Antares and Arcturus, become supernovae, and our solar system makes multiple orbits around the centre of the Milky Way. Local dwarf galaxies such as the Magellanic Clouds will likely be cannibalized by the Milky Way and vanish as separate galactic systems forever.

Data and supercomputer simulations now show that in about four billion years, the Milky Way and the Andromeda Galaxy will collide and merge, forming a massive giant elliptical galaxy. And the dense growing galaxy will temporarily erupt as a Seyfert system as the two supermassive black holes in the current galactic cores merge. The new giant elliptical galaxy, called 'Milkomeda', will dominate the dynamics of the other 54 galaxies in the Local Group, and by about 150 billion years only Milkomeda will survive after multiple other episodes of collision and cannibalism.

Milkomeda

STAR DEATH

Stars eventually grow old and die. Across the visible universe, we see countless elliptical galaxies whose stars are already ancient, with no more interstellar gases and clouds from which to fashion newer generations. Stars with our Sun's mass can last 12–15 billion years before evolving into planetary nebulae and cooling white dwarf stellar cinders. But the smallest stars capable of thermonuclear fusion have only 6 per cent of our Sun's mass and are extremely common. These red dwarf stars can survive between 10 to 20 trillion years before becoming inert, cooling objects. Proxima Centauri, a nearby red dwarf, will reach this stage in about four trillion years. So we can surmise that for our visible universe, after a few trillion more years, even the red dwarf stars will have faded from view and then will at last go dark forever. This is the most conservative outlook on the future of our universe based on what we already know about star formation and evolution in our universe.

RED GIANT ▶ *a luminous giant star in the final stages of stellar evolution, which is in the process of using up the last of its helium fuel and has expanded, emitting light in the red-orange part of the spectrum.*

PLANETARY NEBULA ▶ *a ring-shaped expanding shell of gas expelled by a red giant star.*

WHITE DWARF ▶ *what remains when a red giant has expelled its outer atmosphere as a planetary nebula and shrunk down to a hot white dense core.*

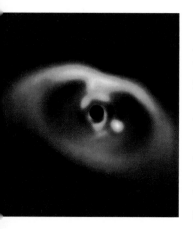

A typical black dwarf remnant of a dead star.

Once the lights go out, there will still be daily flashes of gamma rays and pulses of gravitational radiation passing through the dark spaces as neutron stars and black holes collide following their various gravitational dances. We see this today, and in the far future there will still be plenty of neutron stars and black holes in binary systems to create this spectacle.

But there is potentially another phenomenon that may have even more drastic affects on what we may see. As a consequence, we may not last more than another 65 billion years.

If the dark energy in our universe continues to grow in strength, the accelerating separation of galaxies in space will eventually cause our neighbourhood to thin out dramatically in only 50 billion years. The consequence is that our combined Andromeda-and-Milky-Way galaxy, 'Milkomeda', will become the only source of light in our local universe by 60 billion years from now. Even nearby superclusters of galaxies, such as Virgo at 110 million light years and Coma at 330 million light years, will be so far away, their light will be beyond the 60 billion light year visible universe horizon at that time, and permanently invisible from Earth. Between this horizon and Earth, only a few small galaxy groups currently close to us today will be a part of our visible universe, which can now be numbered as containing dozens or perhaps a few hundred remaining galaxies rather than the tens of billions of galaxies that now make up the extragalactic view.

THE BIG RIP

If dark energy continues to grow in strength without limit, the scale at which the visible universe horizon exists will continue to shrink, first to extragalactic scale at 60 billion years, then accelerating to galactic, solar system, planetary and atomic scales by 65 billion years hence. Some estimates even predict the ripping apart of space itself at the quantum gravity scale.

Ironically, because stars are still plentiful and bright in the precursor to such an event, a night sky, like the one we currently enjoy, may rapidly transform over the course of millennia into an increasingly sparse and redshifted panorama of stars fleeing from our location. Within a human lifespan, a wave of increasing redshifted stars will seem to approach us with nearer stars less redshifted than farther ones that are rapidly becoming dimmer as their spectra shift into the *Spacetime disintegrates during the Big Rip.*

infrared. In the final moments, the distant planets in the solar system will also start to become redshifted until finally this plague of space dilation reaches Earth itself. Following a short period of time when we find ourselves alone in a dark cosmos, the atomic structure of Earth and our own bodies will start to dilate until death overtakes us.

This dismal forecast can be postponed almost indefinitely if the phenomenon responsible for dark energy dissipates by converting the current false vacuum into a slightly lower-energy true vacuum, at which point the accelerating expansion will come to an end over the next few billion years. However, it is not clear what the energy difference between the current true and false vacuum might be. If it is in the order of less than the masses of neutrinos, a few tenths of an electron-Volt, then very little change will occur to the current physics of the universe. However, if the energy difference is comparable to nuclear-scale energies of MeV or

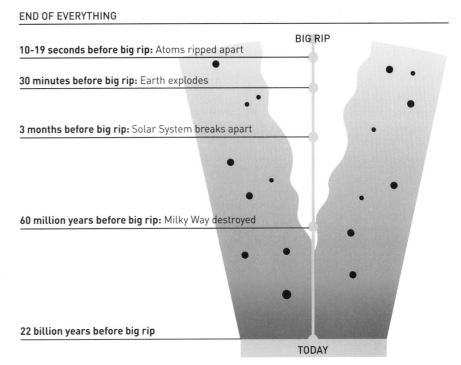

END OF EVERYTHING

BIG RIP

10-19 seconds before big rip: Atoms ripped apart

30 minutes before big rip: Earth explodes

3 months before big rip: Solar System breaks apart

60 million years before big rip: Milky Way destroyed

22 billion years before big rip

TODAY

particle mass energies of GeV, then our universe will be utterly transformed as a new Standard Model replaces the older one. The new Standard Model will look nothing like the one upon which our organic life is based, and so as these bubbles of the true vacuum expand at light speed, they will annihilate any forms of familiar matter that they touch, including Earth, and our form of life throughout the universe.

THE BIG CHILL

If space is not torn asunder in another 65 billion years in the Big Rip, there are other possibilities we can consider. Many speculative calculations have been attempted which lead to an even more barren landscape after vast tracts of time measured in googleplexs (e.g. 10^{100} years) have passed.

Galactic remnants

After all stars have been converted to cold, dense remnants such as black dwarfs, black holes and neutron stars, our galaxy will be a dark graveyard of non-luminous degenerate matter. By about 100 trillion years, occasional collisions between these remnants will cause supernovae. By 10^{20} (or 100 billion billion) years, most stellar remnants (neutron stars, white dwarfs and black holes) will be gravitationally ejected from their galaxies by close collisions. After 10^{30} years, surviving remnants will be absorbed by galactic supermassive black holes through the steady emission of gravitational radiation causing orbit decay.

This degenerate graveyard universe will not be the final stage in conceivable cosmic future history but only an inflection point to an even longer and more peculiar time called The Black Hole Era. By 10^{70} years, all remaining stellar mass black holes will have evaporated by the Hawking mechanism, leaving behind elementary particles like electrons, positrons, photons and neutrinos. Then by 10^{99} years, the most massive supermassive black holes in our universe today will also evaporate by the Hawking mechanism.

Heat death of the universe

Once the last remaining black holes have evaporated, we enter an unimaginably long period of time called the Dark Era. What is left behind at this time is an expanding universe filled with no forms of matter other than a dilute plasma of electrons, positrons, photons and neutrinos, which represents the final 'heat death' of the cosmos. There is no longer any way to generate usable energy in a cosmos where entropy is now at its maximum possible level. If it should turn out that on these timescales even electrons, positrons and neutrinos are not elementary particles, they may decay into their constituents, which may resemble photons in virtually every other property.

ETERNITY

Current measurements of the mass of the top quark and Higgs boson imply that our current vacuum state is unstable. By quantum tunnelling, our current false vacuum state could transform into a true vacuum in about 10^{139} years.

Black Hole Evaporation

In 1974, Stephen Hawking announced his discovery that, due to quantum processes near the event horizon, black holes had an effective surface temperature, which means they were losing energy and therefore mass. The lifetime of a black hole against this quantum evaporation process, called 'Hawking radiation' or the 'Hawking mechanism', is given in seconds for a black hole with a mass of M grams, by

$$t = 4.8 \times 10^{-27} M^3$$

An object like our Sun with $M = 2 \times 10^{33}$ g would take about 4×10^{73} s or about 10^{66} years – far longer than the age of our universe. However, black holes with a mass of 4×10^{14} g would take about 14 billion years. This led astronomers to speculate that if black holes could be created during the Big Bang, those with masses near 10^{14} g would be in their last stage of evaporation, and cause energetic bursts of gamma rays and other radiation. Searches for these events have not turned up any candidates.

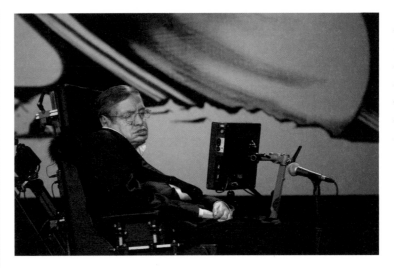

Stephen Hawking found out that black holes were losing mass through a process now known as 'Hawking radiation'.

After $10^{10^{26}}$ years, all particles larger than the Planck mass will quantum tunnel directly into black holes.

After $10^{10^{26^{56}}}$ years, the empty expanding universe will spontaneously create daughter universes by quantum tunnelling. These will go on to have their own Big Bangs.

Daughter universes

THE END

Current cosmological models show that our universe is extremely vast, and may in fact be infinite for all intents and purposes. Big Bang cosmology also describes a cosmos in the state of continued expansion that may continue into the indefinite future. If that is the case, and given what we know about gravity and the Standard Model today, one thing appears to be exceedingly clear. In the vast expanse of the eternity of time to come, our current, life-nurturing, stellariferous universe spans only the first 100 trillion years of eternity, to be followed by a darkness so deep and complete that we humans have no words for it.

The first 100 trillion years

It is indeed sobering to contemplate that this Era of Life is but a brief flash of light in the eternal depths of darkness and times to come.

EARLY IDEAS ABOUT THE FUTURE

Before the current Big Bang model was observationally established to forecast open-ended future expansion, we had various Friedman models that predicted the eventual re-collapse of the cosmos called the Big Crunch. Based on the then best available values for the cosmological parameters of density and the Hubble constant in the 1960s, some calculations suggested we were about two-thirds of the way to the point of maximum expansion, and so the time to the Big Crunch was estimated to be in another 40 billion years or so. The Big Crunch option vanished once accurate measures of the current cosmology were made, starting in the 1990s, and it was found that re-collapse is nearly impossible, especially given the nature of dark energy.

Before the modern age of physical cosmology, beginning in the 1900s, the concept of the distant future was mostly relegated to fiction writers and religious ideas. Hindu religion offered a cyclic universe with a duration of some eight billion years each cycle, while Western religions favoured an end to humanity at the so-called Second Coming, which was apparently to take place in the near future but, again, no theological mention was made of what happens to the rest of the universe.

Among fiction authors, in H.G. Wells' *The Time Machine* (1895) we read about a distant Earth 30 million years from now. It is a desolate planet. But no comment about the universe at large. One of the earliest science fiction stories about the end of the universe can be found in Olaf Stapledon's 1937 book *Star Maker*. This author's favourites are James Blish's story *The Triumph of Time* (1958) and Michael Moorcock's novel *The Sundered Worlds* (1965), which has the first literary use of the term 'Multiverse', which predates significantly its use by modern cosmologists. In most recent science fiction we have far futures developed by Frederick Pohl's *The World at the End of Time* (1990), and Stephen Baxter's breathtaking epic novels *Ring* (1994) and *Manifold:Time* (1999).

Frederick Pohl, a leading science fiction author, often wrote about the far future and explored the concept of the multiverse in his novels.

Chapter Fourteen
TIME

Chronological Concepts – Modern Physics – The Specious Present –
Biological 'Now' – Internal Model of History – Physical 'Now' –
Arrow of Time – Time as a Physical Concept

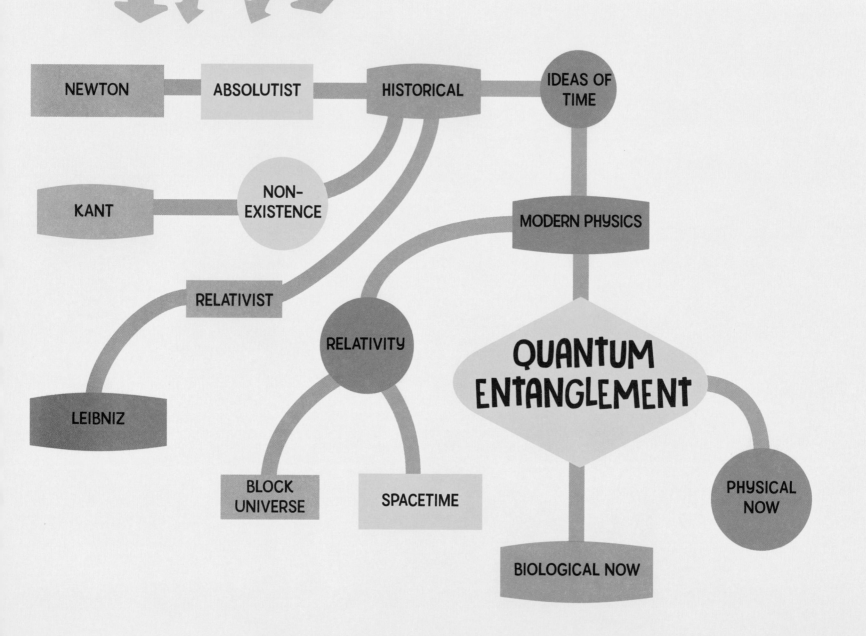

NEWTON — ABSOLUTIST — HISTORICAL — IDEAS OF TIME

KANT — NON-EXISTENCE

MODERN PHYSICS

RELATIVIST

RELATIVITY

QUANTUM ENTANGLEMENT

LEIBNIZ

BLOCK UNIVERSE

SPACETIME

PHYSICAL NOW

BIOLOGICAL NOW

CHRONOLOGICAL CONCEPTS

The entire discussion of cosmology has followed two historical treatments. The first is the human history of discovery, which began with ancient speculation, and has evolved over the millennia to the current, detailed, physical description of Big Bang cosmology. The second history is that of the physical universe itself, from the complex changes during the Planck Era, to the profound physical decay that awaits the universe in the unimaginably distant future. Implicit in both of these descriptions is the idea of events strung together into a causal chain (Event A caused Event B which caused Event C etc.) that is ordered in time.

Although we have described ideas in quantum gravity theory which give our best understanding of the origin and meaning of space, we have not offered any description of the origin of time and why time appears to be a different phenomenon and characteristic of the world than space.

THREE DESCRIPTIONS HAVE APPEARED IN RECENT CENTURIES OF THE ORIGIN OF TIME:

ABSOLUTIST

Sir Isaac Newton proposed that time and space are absolute in that there is a pre-existing external framework for clocks and rulers that functions behind the scenes and allows physical laws and theories to make mathematically precise predictions about dynamical systems.

RELATIVIST

Gottfried Wilhelm Leibniz proposed that there is no Newtonian 'absolute framework'. Both space and time are concepts derived from the relationships between bodies. Without bodies, time and space would not exist.

NON-EXISTENCE

Immanuel Kant (c.1788), in his *Critique of Pure Reason*, proposed that both time and space are merely an a *priori* intuition (that is derived from intellectual reasoning rather than experience) which allows us to organize sensory experiences into a meaningful model of the world within which we can then operate. This view mirrors the idea first proposed by Antiphon the Sophist in the 5th century BCE: '*Time is not a reality* (hypostasis), *but a concept* (noêma) *or a measure* (metron).'

MODERN PHYSICS

Time is one of the oldest mysteries of our world, and humans have struggled with it for millennia, especially the profound sense we have that it flows from past to future. Only the advent of modern physical science has seemingly provided us with new tools with which to explore it.

In physics, time is a mathematical symbol in equations often represented by the letter t. It is a convenient parameter with which to describe how a system of matter and energy change. The first very puzzling feature of time as a physical variable is that all mathematical representations of physical laws or theories show that t is continuous, smooth and infinitely divisible into smaller intervals. These equations are also 'timeless' in that they can be written down on a piece of paper and accurately describe how a system changes from start to finish (based on boundary conditions defined, for instance, at '$t = 0$'). But the equations show this process as 'all at once'. The user has to insert the particular value for t that represents the present moment.

Relativity, however, brought with it a new way of looking at space and time as members of a spacetime continuum threaded by the worldlines of particles and observers in the universe. Within spacetime, all objects were seen from the 'all at once' perspective that eliminated any mention of time at all. This 'Block Universe' was a complete four-dimensional object, which could be cleaved perpendicular to the time axis to reveal where things were in three-dimensional space at a given moment. But there was no prescription for why one would cleave spacetime at one moment rather than at some other moment.

In quantum mechanics, the evolution of a system through space is described for successive time intervals, but there is no distinction made between any time interval other than at the 'beginning' and 'ending' endpoints and the places where it interacts with its environment. Neither relativity nor quantum mechanics offer an explanation for why time exists, nor why there is a universal sense among humans that there is something special about the present moment in all of cosmic history within the Block Universe.

Block universe

Opposite: This rocket launch can be described mathematically, but no moment in time is favoured over another.

THE SPECIOUS PRESENT

If all the equations and relativistic treatments of the physical world are defined by a timeless mathematical statement embodied in the Block Universe model for spacetime, where does the experience of 'now' come from that mathematically causes a specific time, call it $t = t_{now}$, to be singled out from all the other times defining the system? Relativity and modern physics embed time into an eternal and timeless spacetime framework and do not, therefore, offer an explanation for time other than it is a necessary 'fourth dimension' that organizes the changes in physical systems (states) in space. It is a Kantian-like view of time, but founded on measurable external events. There is a sense in which Kant was correct, and that involves how our brains perceive time. This perception of time, however interesting it is as a topic of research, is not the physical time within which events in the cosmos play themselves out.

EINSTEIN'S BLOCK UNIVERSE

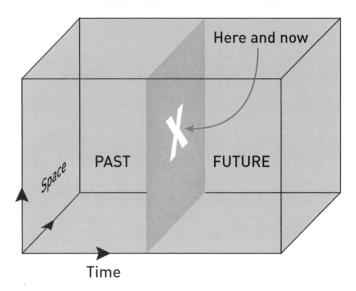

In the relativistic Block Universe view of spacetime, the concept of 'Now' depends entirely on the observer.

BIOLOGICAL 'NOW'

Thanks to sophisticated studies in brain research over the recent years, researchers have identified two to three seconds as the maximum duration of what most people experience as 'now', and what scientists call the 'specious present'. During this time, an enormous amount of neural activity has to occur at nerve cell conduction speeds of up to 10 cm/ms (millisecond). Not only does the sensory information have to be integrated together for every object in your visual field and cross-connected to the other senses, but dozens of specialized brain regions have to be activated or de-activated to update your world model in a consistent way. But this process is not fixed in stone. Sebastian Sauer and his colleagues at the Ludwig-Maximilians-Universität in Munich show that mindfulness meditators can significantly increase their sense of 'now' up to 20 seconds.

Time perception is handled by a highly distributed system involving the cerebral cortex, the cerebellum and the basal ganglia. One particular component, the suprachiasmatic nucleus, is responsible for the circadian (or daily) rhythm, while other cell clusters appear to be capable of shorter-range (ultradian) timekeeping.

INTERNAL MODEL OF HISTORY

The knitting together of these neurological 'nows' into a smooth flow of time is done by our internal model-building system. It works lightning fast to connect one static collection of sensory inputs to another set and hold these both in our conscious 'view' of the world. This gives us a feeling of the passing of one set of conditions smoothly into another set of conditions that make up the next 'Now'. To get from one moment to the next, our brain can play fast and loose with the data and interpolate what it needs. For example, in our visual world, the point where the optic *Brain interpolation* nerve connects with the light-sensitive retina contains no light-sensitive cells and so creates a blind spot in our field of vision, but you never notice it because the brain interpolates across this spot to fill in the scenery. The same thing happens in the time dimension with the help of our internal model to make our jagged perceptions in time into a smooth film-like experience of where a fast-moving ball in flight will be in the next moment.

Neurological conditions such as strokes, or psychotropic chemicals, can disrupt this process. *Neurological* Many schizophrenic patients stop perceiving time as a flow of causally linked events. These *disruptions* defects in time perception may play a part in hallucinations and delusions. There are other milder aberrations that can affect our sense of the flow of time. For example, dischronometria is a condition of cerebellar dysfunction in which an individual cannot accurately estimate the amount of time that has passed.

Because the volume of internal and external data is enormous, we cannot hold many of these consecutive moments in our consciousness with the same clarity, and so earlier 'nows' either pass into short-term memory if they have been tagged with some emotional or survival attributes, or fade quickly into complete forgetfulness. You will not remember the complete sensory experience of diving into a swimming pool, but if you were pushed or injured, you will remember that sequence of moments with remarkable clarity even years later!

The model-building aspect of our brain in time is equivalent to its pattern-recognition ability in space. It looks for patterns in time to find correlations, which it then uses to build up expectations for 'what comes next'. Amazingly, when this feature yields more certainty than the evidence of our imperfect senses, psychologists like Albert Powers at Yale University say that we experience hallucinations. In fact, up to 15 per cent of the human population experiences auditory hallucinations (songs, voices, sounds) at some time in their lives when the brain literally 'hears the sound' because it was strongly expected on the basis of other clues.

PHYSICAL 'NOW'

Philosopher William James defined the specious present to be *'the prototype of all conceived times... the short duration of which we are immediately and incessantly sensible'*. The brain's manipulation of internal model-making and sensory data creates 'now' as a phenomenon we experience, but the physical world outside our collective brain population does not operate through its own neural systems to create a cosmic 'now'. So, in terms of physics, the idea of 'now' does not exist. We even know from relativity that there can be no uniform and simultaneous 'now' spanning large portions of space or the cosmos. Even Einstein himself noted *'...that there is something essential about the Now which is just outside the realm of science.'* If you wanted to define 'now' by a set of simultaneous events spanning space, relativity puts the kibosh on that idea because, due to the relative motions and accelerations of all observers, there can be no simultaneous 'now' that all 'observers' throughout the universe can experience. Also, there is no 'flow of time' because relativity is a theory of worldlines and complete histories of particles embedded in the Block Universe of spacetime. Quantum theory, however, shows us some new possibilities.

Relativity and 'now'

As we have seen in previous chapters, what we call 'space' may be built up like a tapestry from a vast number of quantum events described by quantum gravity. Time may also be created from a synthesis of elementary events occurring at the quantum scale. This is very much like what we call 'temperature'. Temperature is a measure of the average collision energy of a large collection of particles, but cannot be identified as such at the scale of individual particles.

STATES OF QUANTUM PARTICLES

SPIN SPEED POSITION ENERGY MOMENTUM

A system can be described completely by its quantum state, which is a much easier thing to do when you have a dozen atoms than when you have trillions, but the principle is the same. This quantum state describes how the elements of the system are arrayed in three-dimensional space. But because of Heisenberg's Uncertainty Principle, the location of a particle at a given speed is spread out rather than localized to a definite position. A related property of quantum particles is that their states (spin, speed, position, energy, momentum) can become entangled. What this means is that if you prepare two particles close to each other in the same state (called entanglement) in which their quantum properties are correlated with each other and then separate them, their properties will remain correlated over distances far greater than the size of the original system. An intriguing set of papers by physicist Seth Lloyd at Harvard University in 1984 showed that this may be how systems evolve into an equilibrium state. Over time, the quantum states of the member particles become correlated and shared by the larger ensemble. This direction of increasing correlation goes only one way, and establishes the 'Arrow of Time' on the quantum scale.

Quantum entanglement

Quantum Entanglement

When quantum systems interact – such as particle collisions – the interaction causes correlations between the particles involved. They are said to be entangled. This entanglement can be produced in experimental conditions. Once entangled, any measurement made of one particle, such as angular momentum, will give information about the angular momentum of the other, no matter how far apart they are. Einstein called this *'spooky science at a distance.'* But measurement is an active process that alters the system being measured. The process of measuring wipes out all information about other aspects of the particle, such as spin. So measuring the spin of one entangled particle tells you nothing about the angular momentum of the other. Indeed, in quantum mechanics, if a property is not measured it need not even exist.

ARROW OF TIME

Cosmologically, it is apparent from a variety of measures that events are connected in time in a definite past-present-future order. This obtains from a variety of biological, thermodynamic and cosmological measures of closed systems changing their states generally from order to disorder following the Second Law of Thermodynamics. These are called the Arrows of Time, a term popularized by Sir Arthur Eddington in his 1928 book *The Nature of the Physical World*. We always see systems that evolve from an ordered state to a disordered one such as an ice cube melting to water, but we never see water suddenly forming into an organized ice cube.

Because it takes less information to specify an ordered state than a disordered one, physicists see a direct relationship between the entropy of a system and its information content (called its Shannon Entropy). As entropy increases and the system becomes more randomized, the amount of information in the state also increases. Entropy is a measure of the level of disorder in a closed system. The more improbable an event or state, the more entropy and information it contains. These issues of entropy and information content are playing a big role in the study of black holes and quantum gravity theory in which the surface area of a black hole plays the role of its entropy, and is related to the number of bits of information that can be stored on its horizon limited by the Planck scale.

Shannon Entropy

In Loop Quantum Gravity, spin networks in 3-D become spin foams in 4-D.

ENTROPY ▶ *a measure of the level of disorder in a closed system.*

TIME AS A PHYSICAL CONCEPT

Discussions about time itself are not found in Superstring Theory, which implicitly assumes the pre-existence of a special relativistic spacetime within which the strings exist and move. Currently, only within Loop Quantum Gravity, which we discussed in Chapter 12, has there been considerable attention paid to the emergence of time as a physical concept.

As discussed by theoretical physicists Lee Smolin and Carlo Rovelli, spacetime is constructed at the quantum level from what are called spin networks. Like a vast Tinkertoy model, these networks consist of nodes that represent quantized volumes of space at the Planck scale, connected by relationships represented by edges that carry quantized units of area. At this level, space is created at the quantum level from a network of these point volumes and relationship edges. Each of these networks represents a state of space, but these states are part of a four-dimensional network called a spin foam, which represents how the linkages in one spin network rearrange themselves into another spin network along a sequence of changes. Spin foams are the quantum version of spacetime in which the fourth dimension organizing the spin networks is what we identify as time in the larger-scale 'classical' spacetime of relativity.

Spin networks

> *'Science is not about certainty. Science is about finding the most reliable way of thinking...not only is it not certain, but it's the lack of certainty that grounds it.'*
> —Theoretical Physicist Carlo Rovelli

Nevertheless, even within LQG, there is no explanation for why the organization of three-dimensional spin networks in a spin foam is treated as a time-like organization of three-dimensional things rather than merely a four-dimensional space-like structure. This is often called the 'Problem of Time' and it is an outstanding challenge for nearly all versions of quantum gravity theory.

Problem of Time

One clue about time is that the Wheeler-DeWitt equation, which we discussed earlier as the origin of the loop solutions in Loop Quantum Gravity, makes no reference to time as an external variable. Time only appears as an internal variable. This reflects the idea that there is no 'time' outside what we define as the universe, but only something that observers inside a universe experience through the use of various sub-systems called 'clocks'.

Ekaterina Moreva

Ekaterina Moreva was born in Murashi, Russia in 1981. She graduated in Engineering Physics at the Moscow Engineering-Physics Institute (MEPhI), Russia in 2004, and received the PhD degree in Physics and Mathematics at MEPhI, in 2007. From 2005 to 2011 she was an assistant professor at the MEPhI Department of Theoretical and Experimental Physics. Since 2012, she has been a researcher at the Istituto Nazionale di Ricerca Metrologica, Italy, Turin. Her research interests include quantum information, quantum tomography, quantum cryptography, and the foundations of quantum mechanics.

This idea that no external time exists but is something created by internal configurations of matter, was originally proposed in 1983 by theoretical physicists Don Page and William Wootters, and finally put to the test by physicist Ekaterina Moreva at the Instituto Nazionale di Ricerca Metrologica in Italy. The results showed that when the polarization of an entangled system of two photons was measured no changes occurred, because the properties (polarization) of both photons was measured outside the entangled system. This is the perspective of a 'super-observer' in the Wheeler-DeWitt universe. However, when the polarization of one photon was measured from within the system, the observer becomes entangled with it. When this measurement is compared with the polarization of the second photon also within the system, the difference is a measure of time. This confirms that time is an emergent property of an entangled system as perceived by 'clocks' within a universe, but not by an external clock as for the Wheeler-DeWitt system.

Super-observer

Experiments such as these at the quantum level suggest that the Block Universe perspective of relativity is wrong. It is conceptually based on a spacetime diagram with space on the horizontal axis and time on the vertical axis. But this can't be correct because if time is an emergent phenomenon, there is no well-defined time axis to draw vertically in the Block Universe representation. Instead, we are faced with the idea that although the past

Problems with the Block Universe

can be reconstructed from essentially classical physics and records of stored information (photographs, etc), the future is closely determined by probabilities and principles found in quantum mechanics.

The 'present' is when the quantum mechanical probabilities of the future become 'crystallized' into the certainty of the past. Physicist George Ellis in 2009 proposed this as the concept of the Crystallizing Block Universe, which is consistent with what LQG founder Lee Smolin proposes: '*The future is not now real, and there can be no definite facts of the matter about the future. [What is real is] the process by which future events are generated out of present events.*'

NAMES TO KNOW: THE PHYSICS OF TIME

Carlo Rovelli

Don Page

William Wootters

Ekaterina Moreva

George Ellis

In this visual analogy, the roots of a tree represent crystallized past worldlines. The trunk is the present moment, and the canopy of leaves is the quantum indeterminacy of the future.

PHYSICAL TIME AS AN EMERGENT PHENOMENON

The idea that time is not a fundamental property of nature but an emergent one is similar to the concept of temperature, which is well defined for large collections of particles but is meaningless for individual particles, as it is a measure of the average kinetic energy of the particles. Other examples of emergent phenomena include the well-known properties of liquid water, turbulence, air pressure, rainbows, life and even consciousness itself. Closely related to emergence is the idea of self-organization. Opposite we have individual objects (birds) operating collectively by applying a simple rule (stay close to your neighbour; when they move, you move too). As 'murmurations' of starlings show, this can lead to beautiful and complex patterns in time and space.

Another mechanism for the emergence of time is via a process called quantum tunnelling (see pages 191–2). At the atomic and nuclear scale, spontaneous transitions appear in the decay (or *fissioning*) of certain nuclei. For example, a Polonium-212 nucleus spontaneously emits a helium nucleus in a process called alpha decay. The time you have to wait for this for a collection of polonium nuclei is about 0.2 microseconds. The escape of the alpha particle from the polonium nucleus is not permitted classically because it violates the conservation of energy. The alpha particle just doesn't have enough energy to break free from the nucleus, but escape is permitted quantum mechanically because of Heisenberg's Uncertainty Principle. In other words, you can't really tell if the alpha particle at a specific location in the nucleus has exactly the energy it needs to escape. The time needed to escape the nucleus and 'tunnel' through the energy barrier depends on the energy difference. The bigger the energy difference, the longer it will take in an exponential way. A controlled application of quantum tunnelling can be found in the Scanning Tunnelling Microscope, which is able to detect individual atoms in a number of different systems by detecting the electron tunnelling current.

According to Stephen Hawking, a similar tunnelling process may have occurred at the Big Bang. In its initial state, which could have been similar to the spacelike attributes of the spin networks in LQG, cosmic spacetime may have been in a four-dimensional, pure space state. The spin foam we discussed previously may have been a purely four-dimensional space-like object. But then through a tunnelling event, one of the space-like dimensions tunnelled

into a time-like dimension and, quite literally, time began. From then on one might suppose that the quantum entanglement process created events and sub-systems called 'clocks' from which state changes would be interpreted as on-going, time-like changes between states. Taking Hawking's idea one step further, the tunnelling event may have been irregular so that some regions of the pre-existing four-dimensional space may have remained unaffected while other 'bubbles' of the true spacetime may have formed with a time-direction.

A murmuration of starlings shows how complexity emerges from simplicity.

GLOSSARY

1. **Accretion disk** stellar debris, including gas and dust, which has been pulled into a flattened band of spinning matter around a black hole.

2. **Angular momentum** the level of spinning movement in a rotating body. It is 'conserved' (remains the same) unless acted on by another force.

3. **Astronomical unit (AU)** the distance between the Earth and the Sun; a relative scale that compares distances between the Sun and other solar system bodies.

4. **Balmer absorption lines** in spectroscopy, absorption lines seen in the spectrum of hydrogen atoms when viewing stars with a surface temperature of more than 10,000 K.

5. **Baryon** subatomic particle that has a mass equal to or greater than a proton. Baryons belong to the quark-based hadron family of particles.

6. **Block universe** the theory that all objects and events in the universe – past, present and future – are all together in one 4-D block.

7. **Boson** subatomic particle such as a photon with zero or integral spin.

8. **CBR** cosmic background radiation: the light energy produced by the Big Bang. The CBR has cooled over time and is now only detectable at microwave wavelengths, hence cosmic microwave background (CMB) radiation.

9. **Dark energy** a component to the contents of the universe which accounts for its accelerated expansion rate.

10. **Dark matter** a significant contribution to the gravitating mass of the universe not in the form of known types of matter such as electrons, protons, photons or neutrinos.

11. **Doppler effect** The shifting of the wavelength or frequency of sound or light in response to the relative motion between the source and the observer.

12. **Entropy** a measure of the level of disorder in a closed system.

13. **Event horizon** boundary of a black hole, beyond which nothing can escape – not even light.

14. **Field.** the distribution of some quantity throughout space such as radiation intensity, gravity or temperature.

15. **Fine structure constant** a number (close to $1/137$) that relates to the strength of the electromagnetic force and governs how elementary charged particles (electrons and muons) and light (photons) interact.

16. **Geodesic curve (or arc)** the shortest distance between any two points on a curved surface. In *Flatland* (see page 52), a 2-D being living on a sphere and following a geodesic curve would think it had travelled in a straight line, although it had actually described an arc.

17. **Gravitons** the quanta (discrete units or building blocks) of the gravitational field.

18. **Hadron** subatomic particle, such as a baryon or a meson, that can take part in the strong interaction.

19. **Homogeneity** [see also Isotrophy] the principle that matter has a uniform distribution along the third dimension out into the depths of space.

20. **Horizon problem** the theory that all areas of the universe have maintained a uniform temperature despite not being in contact for nearly 14 billion years. Cosmic inflation is the currently accepted explanation for this.

21. **Hubble's law** also known as the Hubble–Lemaître Law, this states that objects in deep space are observed to have a redshift, interpreted as a relative velocity away from Earth.

22. **Inflaton field** a theoretical scalar field that may drive cosmic inflation in the early universe.

23. **Inverse-square law** that intensity (e.g. of brightness) is inversely proportional to the square of the distance from the source; that is, as distance increases intensity decreases using the formula $1/d2$.

24. **Isotrophy** from the standpoint of any observer, the principle that matter seen across the two-dimensional sky looks uniform in angular degree.

25. **Lepton** elementary particle that contains one unit of electric charge and responds only to the electromagnetic force, gravitational force and weak force – not the strong force. They include electrons, muons, tauons, neutrinos and their antimatter variants.

26. **Loop quantum gravity** a theory of gravity in which 4-dimensional spacetime is fashioned from loops of relationships whose interlinkages give rise to space and time.

27. **Mass-to-light ratio** a figure derived by dividing the mass of a star, galaxy or cluster by its luminosity. By identifying a star type and measuring its luminosity, you can work out its mass.

28. **Meson** subatomic particle, intermediate in mass between an electron and neutron, composed of one quark and one anti-quark. It transmits the strong interaction that holds nucleons together in the nucleus.

29. **Neutron star** a very small, very dense star resulting from the gravitational collapse of a massive star after a supernova explosion.

30. **Nucleosynthesis** the formation of new atomic nuclei from protons and neutrons.

31. **Parallax** the apparent shift in the position of an object in space when viewed from two different locations. Using geometry, the amount of this shift can be used to calculate the distance to the object.

32. **Planck era** the very earliest stage in the birth of the universe, immediately after the Big Bang.

33. **Planetary nebula** a ring-shaped expanding shell of gas expelled by a red giant star.

34. **Proper motion** speed and direction in two dimensions, as viewed from earth.

35. **Pulsar** rapidly rotating neutron star or white dwarf that emits a very powerful beam of electromagnetic radiation. It was discovered by radio astronomer Jocelyn Bell Burnell.

36. **Quantum** (plural **quanta**), discrete packets of subatomic energy; the smallest amount of energy that can be involved in an interaction.

37. **Quantum chromodynamics** a quantum field theory in which the strong interaction is described in terms of an interaction between quarks mediated by gluons. Quarks and gluons are both assigned a quantum number called 'colour'.

38. **Quantum electrodynamics** field theory that unites quantum mechanics and Special Relativity to explain how light and matter interact.

39. **Quantum fluctuations** following from Heisenberg's Uncertainty Principle, temporary changes in the amount of energy/appearance of energetic particles out of nothing. It allows for particle-antiparticle pairs of virtual particles to form.

40. **Quantum gravity** The idea that spacetime and the gravitational field are quantized at the Planck scale, much like the electromagnetic field is comprised of individual photons.

41. **Quantum tunnelling** in quantum mechanics, when a particle such as an electron passes through a barrier that, in classical physics, should repel it. According to the Heisenberg Uncertainty Principle, the particle has some finite probability of overcoming the barrier, known as 'tunnelling'.

42. **Quark** elementary particle; quarks combine to form composite particles such as hadrons. There are six types: Up, Down, Top, Bottom, Strange and Charmed, plus their anti-matter variants.

43. **Quasars,** or 'quasi stellar radio sources', are massive sources of radio energy found in the nucleus of remote galaxies. They may contain massive black holes.

44. **Radial velocity** speed away from (or towards) the Earth, as measured by Doppler shift.

45. **Red giant** a luminous giant star in the final stages of stellar evolution, which is in the process of using up the last of its helium fuel and has expanded, emitting light in the red-orange part of the spectrum.

46. **Redshift** the displacement to longer wavelengths of the light from distant galaxies due to the expansion of space.

47. **Relativity** the principle that the properties of objects and their motions are defined only in terms of relational principles, not the absolute properties of space and time.

48. **Scalar** a quantity in physics that has magnitude only, such as 10 kg, 20 cm etc, and no other characteristic.

49. **Scalar field** a scalar quantity applied to every point in a given space, such as the background temperature of the universe.

50. **Space** a property of spacetime derived from the relations between the worldlines of particles, rather than the properties of an absolute pre-existing 3-dimensional framework.

51. **Space velocity** speed and direction in three-dimensional space.

52. **Spin** a property of an elementary particle analogous to rotational movement (intrinsic angular momentum), which confers magnetic field and electric charge in macro molecules. It is not a perfect analogy because the particle does not actually spin!

53. **Standard model** the current comprehensive theory describing the known forces and elementary particles except gravity.

54. **String theory** a description of particles in terms of loops of 'string' vibrating in up to 10 dimensions while travelling through a background spacetime.

55. **Strong force (or Interaction)** one of four fundamental forces governing all matter; it binds quarks together to form composite particles such as baryons (e.g. protons and neutrons).

56. **Supersymmetry** the principle that proposes a relationship between fermions and bosons and that each particle in one group has a paired 'superpartner' in another group. It is intended to fill in gaps and inconsistences in the Standard Model.

57. **Symmetry** an aspect of a system that stays the same after some transformation. Knowing that some parts stay the same facilitates the discovery of the unknowns in the system.

58. **Time** A feature of spacetime in which the changing properties of objects are measured by devices called clocks rather than rulers.

59. **Vacuum energy** the underlying energy of space that follows from Heisenberg's energy--time uncertainty principle.

60. **Weak interaction** one of four fundamental forces governing all matter; it works at short distances between subatomic particles and causes radioactive decay. In weak reactions, particles may disappear or reappear.

61. **White dwarf** what remains when a red giant has expelled its outer atmosphere as a planetary nebula and shrunk down to a hot white dense core.

62. **Worldline** the path that an object takes in 4-D spacetime; it includes the object's position at different moments in time from the past to the future as well as its location within 3-D space.

63. **Zero-point energy** in quantum mechanics, the lowest possible energy of a physical system at its ground state. It is still more energy than allowed for in classical physics because of the Uncertainty Principle.

64. **Zodiac** a belt of sky that encompasses the apparent paths of the Sun, Moon and visible planets. It is divided into 12 regions, each one named after the constellation it holds.

SUGGESTED READING

1: DISCOVERING THE UNIVERSE

Mithen, S. (1997). *The Prehistory of the Mind*. Thames and Hudson Publishing.

An introduction to how the human mind evolved step by step from the cognitive abilities of primitive mammals. It discusses the brain regions and circuits that must pre-exist before our conscious deliberation of the world can arise and sets the stage for understanding how scientific thinking is an extension of our evolved model-building experience.

Sachs, O. (1985). *The Man who Mistook His Wife for a Hat*. Summit Books.

How we investigate the world to create a cosmological theory requires that our brains be organized in a specific way to perceive relationships. This process goes wrong with people suffering after a stroke from a condition called anosognosia. Sachs details the many ways that our brains can fabricate realities and get us to believe in very odd sensory worlds.

CHAPTER 2: COMPOSITION OF THE UNIVERSE

Rees, M. (1997). *Before the Beginning: Our Universe and Others*. Helix Books.

Rees, an astrophysicist, discusses many issues in cosmology from the standpoint of astronomical research, including observational Big Bang cosmology, Einstein's general relativity, the origin of galaxies, and speculations about the origin of the universe.

CHAPTER 3: THE RELATIVITY REVOLUTION: SPACE 2.0

Einstein, A. (1922) *The Meaning of Relativity*. Princeton University Press.

This excellent guide to relativity described space, time and spacetime in detail and how to think about them within the context of relativity. Included are his comments that space is a fiction.

CHAPTER 4: RELATIVISTIC COSMOLOGY

Hawking, S. (1988) *A Brief History of Time*. Bantam Books.

This theoretical physicist has been at the forefront of gravitation research and cosmology since the 1970s. This is his award-winning book that outlines for the general public what we know about the universe and the Big Bang, and what remains to be understood through a scientific approach.

CHAPTER 5: THE DARK UNIVERSE

Nicolson, I. (2007) *Dark Side of the Universe: Dark Matter, Dark Energy, and the Fate of the Cosmos*. Johns Hopkins University Press.

Covers the discovery of dark matter and dark energy through astronomical investigations and how physicists are searching for its causes in the Standard Model and beyond.

Freeman, K. and McNamara, G. (2006) *In Search of Dark Matter*. Springer-Praxis Books

This is a wide-ranging guide to dark matter and how it arose in astronomy based on observations in physics and astronomy. Written for the general public in a mostly non-technical style.

CHAPTER 6: WHAT IS MATTER?

Oerter, R. (2006) *The Theory of Almost Everything: The Standard Model, the Unsung Triumph of Modern Physics*. Pearson Education Press.

Describes the details of the discovery of the Standard Model and how it forms the basis of the deep structure of the universe.

CHAPTER 7: BEYOND THE STANDARD MODEL

Pagels, H. (1985). *Perfect Symmetry: The Search for the Beginning of Time*. Simon and Schuster.

Describes the entire issue of symmetry in physics, including the search for a Grand Unified theory. He discusses the issue of cosmic origins, Planck-scale physics, black holes and inflationary cosmology in a very popular writing style.

CHAPTER 8: THE BEWILDERING GALAXY ZOO

Dickinson, T. (2017) *Hubble's Universe: Greatest Discoveries and Latest Images*. Firefly Books.

This is a comprehensive guide to the discoveries by the Hubble Space Telescope, including its deep space studies of quasars and infant galaxies.

Tyson, N. (2017) *Astrophysics for People in a Hurry*. W.W. Norton and Company.

A delightful and fast-paced introduction to the basic objects in the universe that form the ingredients to cosmology.

CHAPTER 9: THE FIRST STARS AND GALAXIES

Mather, J. and Boslough, J. (1996). *The Very First Light: The True Inside Story of the Scientific Journey Back to the Dawn of the Universe*. Basic Books.

Mather was the Principle Investigator for the COBE instrument that determined the temperature of the cosmic microwave background and its black body spectrum. This book describes the history of creating such an instrument and getting NASA to approve the COBE spacecraft for launch in 1988.

CHAPTER 10: ORIGIN OF THE PRIMORDIAL ELEMENTS

Smoot, G. (1994). *Wrinkles in Time: Witness to the Birth of the Universe*. William Morrow & Company.

Smoot was the Principle Investigator for the COBE satellite instrument that discovered the anisotropy in the cosmic microwave background radiation. This book details the path he took to the Nobel Prize that he shared in 2006 with John Mather.

Weinberg, S. (1977). *The First Three Minutes*. Basic Books.

This is a must-read book that describes the conditions after the end of the Lepton Era. His pioneering investigation of the first three minutes after the Big Bang legitimized this area of cosmological research. This book describes his thoughts about the origin of the universe and its evolution.

CHAPTER 11: INFLATIONARY COSMOLOGY

Greene, B. (1999) *The Elegant Universe: Superstrings, Hidden Dimensions, and the Quest for the Ultimate Theory*. W.W. Norton and Co.

This book is an extremely well written guide to string theory. The issues of hidden dimensions and the multiverse are explored by a leading string theorist and science popularizer.

Odenwald, S. (2015). *Exploring Quantum Space*. CreateSpace Publishing.

This book covers the many ways that physicists have thought about space, and how current discussions are mathematical models and not likely to be snapshots of what spacetime actually 'look' like. The discussions on string theory and loop quantum gravity emphasize the different objects used to represent these extreme conditions for spacetime.

Weinberg, S. (1992). *Dreams of a Final Theory*. Pantheon Books.

Weinberg details the search for a unified theory of physics that includes gravity. He discusses in detail the missing pieces of the puzzle which would take us beyond the Standard Model. He also discusses the limits to these investigations.

CHAPTER 12: COSMOGENESIS

Barrow, J. & Tippler, F. (1986). *The Anthropic Cosmological Principle*. Oxford University Press.

This book describes the Anthropic Cosmological Principle and why it has come up in modern cosmology, as it considers the initial conditions of the Big Bang. It is semi-technical, and suitable for a college-level introduction to the subject.

Davies, P. (1992). *The Mind of God: The Scientific Basis for a Rational World*. Simon and Schuster.

This acclaimed work of popular science takes on the question of whether there is a supernatural reason for why the universe appears as it does, or if we are just lucky enough to be living within the right kind of cosmos out of all the possibilities. A good companion guide to the Anthropic Cosmological Principle.

Randall, L. (2005). *Warped Passages: Unraveling the Mysteries of the Universe's Hidden Dimensions*. Ecco Press.

This book is a detailed description of string theory and how it describes the nature of matter at the Planck scale. She discusses the string landscape and brane theory and how these portray the nature of spacetime and matter within the 10-dimensional Bulk.

Smolin, L. (2001) *Three Roads to Quantum Gravity: A New Understanding of Space, Time and the Universe*. Basic Books.

A detailed but non-technical guide to the elements of quantum gravity and how string theory and loop quantum gravity are attempting to unravel the details of how spacetime can be quantized. An especially good discussion of background-dependent and background-independent issues in relativity and quantum theory. His description of the universe as processes not things is brilliant, as is his step-by-step guide to loop quantum gravity.

CHAPTER 13: THE FAR FUTURE

Penrose, R. (2011) *Cycles of Time: An Extraordinary New View of the Universe*. Random House.

The best-selling author of *The Emperor's New Mind* and *The Road to Reality*, Penrose provides new views on three of cosmology's most profound questions: What, if anything, came before the Big Bang? What is the source of order in our universe? What is its ultimate future?

CHAPTER 14: TIME

Krauss, L. (2012). *A Universe from Nothing: Why There Is Something Rather Than Nothing*. Atria Books.

This is a highly readable introduction to one of the most curious aspects of the universe – namely, that it exists at all. The discussions range from the discovery of the quantum vacuum to contemporary issues in quantum cosmology.

Roveli, C. (2016). *Reality Is Not What It Seems: The Journey to Quantum Gravity*. Penguin Random House.

This book discusses how time must play a key role in quantum gravity theory but its role is currently hidden and difficult to define. The development of loop quantum gravity is al so discussed through many excellent discussions and the use of analogies.

Smolin, L. (2013). *Time Reborn*. Mariner Books.

This is a companion to his *Three Roads to Quantum Gravity* book that treats how physics has eliminated time as a variable by considering the resulting mathematical equations in a timeless context. Issues such as the origin of the 'now' experience and how this is also missing from physics provide a comprehensive treatment of the 'problem' of time in modern physics.

INDEX